MatWerk

Die inhaltliche Zielsetzung der Reihe ist es, das Fachgebiet „Materialwissenschaft und Werkstofftechnik" (kurz MatWerk) durch hervorragende Forschungsergebnisse bestmöglich abzubilden. Dabei versteht sich die Materialwissenschaft und Werkstofftechnik als Schlüsseldisziplin, die eine Vielzahl von Lösungen für gesellschaftlich relevante Herausforderungen bereitstellt, namentlich in den großen Zukunftsfeldern Energie, Klima- und Umweltschutz, Ressourcenschonung, Mobilität, Gesundheit, Sicherheit oder Kommunikation.

Die aus der Materialwissenschaft gewonnenen Erkenntnisse ermöglichen die Herstellung technischer Werkstoffe mit neuen oder verbesserten Eigenschaften. Die Eigenschaften eines Bauteils sind von der Werkstoffauswahl, von der konstruktiven Gestaltung des Bauteils, dem Herstellungsprozess und den betrieblichen Beanspruchungen im Einsatz abhängig. Dies schließt den gesamten Lebenszyklus von Bauteilen bis zum Recycling oder zur stofflichen Weiterverwertung ein. Auch die Entwicklung völlig neuer Herstellungsverfahren zählt dazu.

Ohne diese stetigen Forschungsergebnisse wäre ein kontinuierlicher Fortschritt zum Beispiel im Maschinenbau, im Automobilbau, in der Luftfahrtindustrie, in der chemischen Industrie, in Medizintechnik, in der Energietechnik, im Umweltschutz usw. nicht denkbar. Daher werden in der Reihe nur ausgewählte Dissertationen, Habilitationen und Sammelbände veröffentlicht. Ein Beirat aus namhaften Wissenschaftlern und Praktikern steht für die geprüfte Qualität der Ergebnisse. Die Reihe steht sowohl Nachwuchswissenschaftlern als auch etablierten Ingenieurwissenschaftlern offen.

Christian Rockenhäuser

Electron Microscopical Investigation of Interdiffusion and Phase Formation at Gd$_2$O$_3$|CeO$_{2-}$ and Sm$_2$O$_3$|CeO$_{2-}$ Interfaces

With a Foreword by Prof. Dr. Dagmar Gerthsen

 Springer

Christian Rockenhäuser
Karlsruhe, Germany

Dissertation, Karlsruhe Institute of Technology (KIT), 2014

MatWerk
ISBN 978-3-658-08792-0   ISBN 978-3-658-08793-7 (eBook)
DOI 10.1007/978-3-658-08793-7

Library of Congress Control Number: 2015933629

Springer

Printed on acid-free paper

Springer is a brand of Springer Fachmedien Wiesbaden
Springer Fachmedien Wiesbaden is part of Springer Science+Business Media
(www.springer.com)

Electron Microscopical Investigation of Interdiffusion and Phase Formation at Gd_2O_3/CeO_2- and Sm_2O_3/CeO_2-Interfaces

Zur Erlangung des akademischen Grades eines
DOKTORS DER NATURWISSENSCHAFTEN
von der Fakultät für Physik
des Karlsruher Instituts für Technologie (KIT)

genehmigte
DISSERTATION
von

Dipl.-Phys. Christian Rockenhäuser

aus Neuenbürg

Tag der mündlichen Prüfung: 27.06.2014

Referentin: Prof. Dr. D. Gerthsen
Korreferent: Prof. Dr. M. J. Hoffmann

angefertigt am
Laboratorium für Elektronenmikroskopie
Karlsruher Institut für Technologie (KIT)

Foreword

Materials with high oxygen-ion and negligible electron conductivity are needed for future high-efficiency Solid Oxide Fuel Cells (SOFCs). State-of-the-art SOFCs are still limited in their performance by material properties, which requires SOFC operation temperatures around 800 °C. High temperatures are, on the one hand, indispensably needed for sufficient oxygen-ion conductivity in the solid electrolyte. On the other hand, degradation of efficiency is an undesirable consequence of the high operation temperatures which is induced by secondary phase formation in the solid electrolyte and electrodes and at interfaces in between. Many problems could be solved in so-called intermediate-temperature SOFCs where the operation temperatures are lowered to 500 to 700 °C. Up to now, Y_2O_3-doped ZrO_2 is the state-of-the-art SOFC electrolyte material but the oxygen-ion conductivity in this material becomes too low with decreasing temperature. Possible alternative material systems are CeO_2-based electrolytes. Especially CeO_2, doped with Gd_2O_3 or Sm_2O_3 seem to be promising, because they are characterized by a higher oxygen-ion conductivity compared to Y_2O_3-doped ZrO_2. The oxygen-ion conductivity of these materials has already been thoroughly studied, but other basic material properties have yet not been investigated for these rare-earth oxide systems.

The PhD thesis of Dr. Christian Rockenhäuser focuses on two aspects. First, phase formation was studied in the $Gd_xCe_{1-x}O_{2-x/2}$ and $Sm_xCe_{1-x}O_{2-x/2}$ systems at temperatures below 1200 °C. Based on these studies conclusions were drawn with respect to the suitability of these materials as SOFC components. Second, cation interdiffusion coefficients were determined and activation enthalpies for Ce/Gd- and Ce/Sm-cation interdiffusion were derived. Thin-film couples were used throughout the work, which were investigated by analytical Transmission Electron Microscopy (TEM). The book gives a thorough literature overview on the $Gd_xCe_{1-x}O_{2-x/2}$ and $Sm_xCe_{1-x}O_{2-x/2}$ phase diagrams. It outlines the basics of diffusion and the peculiarities of phase formation in thin-film couples. An introduction of different TEM techniques and analytical methods is given, which were applied in this work. Particular emphasis was laid on the quantification of composition profiles by Energy-Dispersive X-ray Spectroscopy (EDXS) and Electron Energy Loss Spectroscopy (EELS) performed with high spatial resolution in a transmission electron microscope. The work addresses advanced students and experienced scientists and engineers in basic and applied materials science.

Prof. Dr. Dagmar Gerthsen
Laboratory for Electron Microscopy
Karlsruhe Institute of Technology (KIT)

Acknowledgement

I want to thank Everyone, who contributed to the successful completion of my dissertation.

First and foremost I express my gratitude to Prof. Gerthsen for the excellent supervision and that she made this work possible at the LEM. Her patient revision of my manuscripts helped me a great deal – formulating scientifically correct and precise does not only require practice, but also instruction on "how on does it". I also want to thank to be given the possibility to participate in conferences.

I thank Prof. Hoffmann for the giving the second assessment and the friendly help and support at the IAM-KM with the fabrication of the CeO_2 substrates and targets for the PLD.

I also thank Nicole Schichtel and Prof. Janek at the Physikalisch-Chemischen Institut for the thin film deposition utilizing PLD.

I am grateful to Benjamin Butz for giving the initial impulse to this study. Additionally I was introduced into operating electron microscopes by him and thank him for many productive hours at the microscope.

Many thanks go to Heike Störmer for the correction of several manuscripts and parts of the thesis. And also the discussions about phase diagrams helped me very much.

I also want to give a big thank you to Miss Sauter, who always had some encouragin words for me in every situation.

I also want to express kind regards to my colleagues at the LEM. The discussion of various physical, non-physical and meta-physical topics in the "Kaffeeraum"(coffee room) were very enriching as well as the exquisite pastrys of the "Backstubenbäcker". The LEM-Lauffreunde are also very important, since they see to it that one does not only sit at the desk.

Many thanks to my mom and dad, Florian, Lisa-Marie, Annerose, Wolfgang, and Heike for motivation and support when I needed it.

Stefan, without you I would not succeeded! Thank you, that you were there for me!

Dr. Christian Rockenhäuser

Contents

Foreword   VII

Acknowledgement   IX

List of Figures   XV

List of Tables   XVII

Acronyms   XIX

Symbols   XXI

Introduction   1

1 Fundamentals   3
   1.1 Application of Gd_2O_3-doped CeO_2 (GDC) and Sm_2O_3-doped CeO_2
        (SDC) in Electrochemical Devices   5
   1.2 Phases in Doped CeO_2 Systems   6
        1.2.1 Pure CeO_2   7
        1.2.2 The $Gd_xCe_{1-x}O_{2-x/2}$ System   9
               Pure Gd_2O_3   9
               The $Gd_xCe_{1-x}O_{2-x/2}$ Phase Diagram   12
        1.2.3 The $Sm_xCe_{1-x}O_{2-x/2}$ System   12
               Pure Sm_2O_3   13
               The $Sm_xCe_{1-x}O_{2-x/2}$ Phase Diagram   16
   1.3 Binary Diffusion Couples   17
        1.3.1 Solution of the Diffusion Equation   17
        1.3.2 Binary Diffusion Couples in Multiphase Binary Systems   20
               Qualitative Description   21
               Reaction-Controlled Layer Growth   21
   1.4 Temperature Dependence of Diffusion   24
        1.4.1 The Arrhenius Relation   24
        1.4.2 Grain Growth Experiments   24

2 Experimental Techniques and Instrumentation **29**
    2.1 Electron Microscopical Instrumentation 29
    2.2 Transmission Electron Microscopy (TEM).................................... 30
        2.2.1 Bright-Field Transmission Electron Microscopy (BFTEM). 31
        2.2.2 High-Resolution Transmission Electron Microscopy
               (HRTEM) 31
        2.2.3 Selected Area Electron Diffraction (SAED) 32
    2.3 Scanning Transmission Electron Microscopy (STEM) and Beam
         Broadening... 33
    2.4 Quantitative Energy-Dispersive X-ray Spectroscopy (EDXS)......... 35
    2.5 Electron Energy Loss Spectroscopy (EELS) 38
        2.5.1 Thickness Determination... 39
        2.5.2 EELS Composition Analysis.. 41
               Background Correction... 41
               Composition Quantification.. 42

3 The $Gd_xCe_{1-x}O_{2-x/2}$ System: Phase Formation and Cation
   Interdiffusion **47**
    3.1 Specimen Fabrication... 47
        3.1.1 Substrate Preparation 47
        3.1.2 Thin Film Deposition 49
        3.1.3 Thermal Treatments... 49
        3.1.4 Sample Preparation for Transmission Electron Microscopy 50
    3.2 Microstructure Characterization 50
    3.3 Cation Interdiffusion 54
    3.4 Discussion 56
        3.4.1 Phase Formation in the $Gd_xCe_{1-x}O_{2-x/2}$ System 57
        3.4.2 Interdiffusion at the Gd_2O_3/CeO_2 Interface...................... 59

4 The $Sm_xCe_{1-x}O_{2-x/2}$ System: Phase Formation and Cation
   Interdiffusion **63**
    4.1 Specimen Fabrication... 63
        4.1.1 Substrate Preparation 63
        4.1.2 Thin Film Deposition 63
        4.1.3 Thermal Treatments... 64
        4.1.4 Sample Preparation for Transmission Electron Microscopy 64
    4.2 Microstructural Characterization.. 64
    4.3 Cation Interdiffusion 72
    4.4 Discussion 73
        4.4.1 Phase Formation in the $Sm_xCe_{1-x}O_{2-x/2}$ System 74
        4.4.2 Cation Interdiffusion... 77

4.5   Comparison of the Phase Evolution of the Sm_2O_3-doped CeO_2
       (SDC) and Gd_2O_3-doped CeO_2 (GDC) systems............................ 78

5 Summary 81

Bibliography 85

Own Publications and Contributions to Conferences 101

List of Figures

1.1 Phase diagram of the $Gd_xCe_{1-x}O_{2-x/2}$ system 11

1.2 Phase diagram of the $Sm_xCe_{1-x}O_{2-x/2}$ system 15

1.3 Diffusion layer structure (Distance versus concentration / Concentration versus distance) 22

2.1 Beam spread for CeO_2, Gd_2O_3, and Sm_2O_3 34

2.2 Low-loss spectrum of Sm-1170 40

2.3 Electron energy-loss spectrum from the RL1 of Sm-1170 43

3.1 Scanning Electron Microscopy (SEM) image of densified CeO_2 . 48

3.2 BFTEM images all Gd samples 51

3.3 High-Resolution Transmission Electron Microscopy (HRTEM) image of Gd-1122 52

3.4 Diffractograms and simulated diffraction patterns of Gd-1122 . . 53

3.5 Diffractograms and simulated diffraction patterns of Gd-986, Gd-1069, and Gd-1175 55

3.6 Interdiffusion profiles at the Gd_2O_3/CeO_2 interface 56

3.7 Arrhenius plot of the interdiffusion coefficients of the Gd-samples 60

4.1 BFTEM images of all Sm samples 65

4.2 HRTEM image of Sm-1070 67

4.3 Diffractograms and simulated diffraction patterns of Sm-987, Sm-1175-20h, and Sm-1175-50h 68

4.4 SAED patterns of Sm-1175-50h, Sm-1175, and Sm-1219 70

4.5 Sm-concentration profiles profiles across the Reaction Volumes (RVs), Reaction Layer 1 (RL1), and Reaction Layer 2 (RL2) of the different Sm samples 71

4.6 Interdiffusion profiles at the Sm_2O_3/CeO_2 interface 73

4.7 Arrhenius plot of the interdiffusion coefficients of the Sm-samples 78

List of Tables

1.1 Lattice parameters and atomic positions of CeO_2 and Sm_2O_3 . .   8

2.1 Material parameters for beam spread determination   35
2.2 X-ray emission lines of Ce, Gd, and Sm   36
2.3 Mean concentration ratio determined by EDXS and mean intensity ratios determined by EELS from linescans across Reaction Layers (RLs) and RVs of Sm samples   44

3.1 Denotation of the heat-treated Gd-samples   49

4.1 Denotation of the heat-treated Sm samples   64

Acronyms

BFTEM Bright-Field Transmission Electron Microscopy

CCD Charge-Coupled Device

CIP Cold Isostatic Pressing

EDXS Energy-Dispersive X-ray Spectroscopy

EELS Electron Energy Loss Spectroscopy

FEG Field Emission Gun

GDC Gd_2O_3-doped CeO_2

GIF Gatan Imaging Filter

HAADF High-Angle Annular Dark-Field

HRTEM High-Resolution Transmission Electron Microscopy

HIP Hot Isostatic Pressing

IUCr International Union of Crystallography

IUPAC International Union of Pure and Applied Chemistry

JEMS Java Electron Microscopy Simulations

KIT Karlsruher Institut für Technologie

MIEC Mixed Ionic-Electronic Conductor

PEELS Parallel-Collection Electron Energy Loss Spectroscopy

PLD Pulsed Laser Deposition

RE Rare Earth

RL Reaction Layer (at the Gd_2O_3/CeO_2 interface)

RL1 Reaction Layer 1 (at the Sm_2O_3/CeO_2 interface)

RL2 Reaction Layer 2 (at the Sm_2O_3/CeO_2 interface)

RV Reaction Volume (at the Sm_2O_3/CeO_2 interface)

SDC Sm_2O_3-doped CeO_2

SEM Scanning Electron Microscopy

SAED Selected Area Electron Diffraction

SOFC Solid Oxide Fuel Cell

STEM Scanning Transmission Electron Microscopy

TEM Transmission Electron Microscopy

TIA TEM Imaging and Analysis

TREO Total Rare Earth Oxides

USA United States of America

XRD X-Ray Diffraction

YDZ Y_2O_3-doped ZrO_2

YSZ Yttria-stabilized Zirconia

ZLP Zero-loss Peak

Symbols

A	pre-exponential factor
Al	Aluminum
Ar	Argon
a_i	atomic fraction of the ith atomic constituent
B	fit parameter
b	beam spread
∇	Nabla operator
C	Carbon
$C(x,t)$	concentration dependent on position and time
C_0	initial concentration
C_A	concentration of element A
C_B	concentration of element B
C_s	spherical aberration
C_{sf}	constant surface concentration
$C_{\zeta(mon)}^{eq}$	saturation concentration
C_η	concentration in the layer
$C_{\eta(bix)1}$	concentration at the interface
$C_{\eta(bix)1}^{eq}$	equilibrium concentration at the interface
$C_{\eta(bix)2}$	concentration at the interface
$C_{\eta(bix)2}^{eq}$	equilibrium concentration at the interface
$C_{\theta(flu)}^{eq}$	saturation concentration
Ca	Calcium
Ce	Cerium
Cl	Chlorine
Co	Cobalt
c_1, c_2, c_3	lattice parameters
D	concentration independent diffusion coefficient
$\tilde{D}$	constant interdiffusion diffusion coefficient
$D(C)$	concentration dependent diffusion coefficient
D_0	frequency factor
D_b	grain boundary diffusion
D_V	Volume diffusion
Dy	Dysprosium

d	Fit parameter for the binary diffusion couple solution
E_0	primary electron energy
E_m	average energy loss
Er	Erbium
Eu	Europium
F	relativistic factor
F	Fluorine
Fe	Iron (from Latin: Ferrum)
G	average grain size
G_0	initial grain size
Gd	Gadolinium
Ge	Germanium
H	Hydrogen
Ho	Holmium
I_A	peak intensity of characteristic X-ray line of element A
I_{att}	attenuated photon intesity
I_B	peak intensity of characteristic X-ray line of element B
I_{inc}	incident photon intesity
$\vec{J}$	diffusion flux
J_A	electron intensity of element A
J_B	electron intensity of element B
J_{Back}	background electron intensity
J_{tot}	total electron intensity
J_{ZLP}	electron intensity of the Zero-loss Peak (ZLP)
K	constant depending on the composition of the phase involved in layer growth
K	Potassium (from Latin: Kalium)
k	reaction rate
$k_{A,B}$	Cliff-Lorimer factor
k_B	Boltzmann constant
k_n	material constant for non-parabolic grain growth
k_n^0	pre-exponential factor
k_p	material constant for parabolic grain growth
L	sample length
$L_{i,j}$	ionization edges
$L_{\alpha i, \beta i}$	X-ray emission lines
La	Lanthanum
Lu	Lutetium
l	mass-per-unit area or area density
M	grain boundary mobility

$M_{i,j}$	ionization edges
Mn	Manganese
N	number of diffusing atoms per unit area
N	Nitrogen
N_A	number of atoms per unit area of element A
N_B	number of atoms per unit area of element B
N_V	number of atoms/m^3
Na	Sodium (from Latin: Natrium)
Nd	Neodymium
n	grain growth exponent
O	Oxygen
Pr	Praseodymium
Pm	Promethium
Q	activation enthalpy for the grain growth process
r	fit parameter
Sm	Samarium
Sc	Scandium
Si	Silicon
Sn	Tin (from Latin: Stannum)
s	sensitivity factor for electron energy loss spectroscopy (EELS)
T	temperature
t	time
t_{bix}	thickness of the Reaction Layer (RL) with bixbyite structure
t_e	exposure time
t_l	layer thickness
t_s	sample thickness
Tb	Terbium
Tm	Thulium
$\frac{t_s}{\lambda}$	relative sample thickness
u	spatial frequency
w_i	weight fraction of the ith atomic constituent
x	position coordinate
x_{bf}	interface position
x_{bix}	layer thickness
x_{bix}^*	changeover thickness
x_{mb}	interface position
Y	Yttrium
y	position coordinate
Yb	Ytterbium

Z	atomic number
$\overline{Z}$	mean atomic number
Z_{eff}	effective atomic number
Z_i	atomic number of the ith element in the periodic table
Zn	Zinc
Zr	Zirconia
z	position coordinate
α	angle between basic vectors
β	collection semi-angle
γ	grain boundary energy
Δ	Laplace operator
Δ_A	energy-loss integration window of element A
Δ_B	energy-loss integration window of element B
Δf	defocus
ΔH	activation enthalpy
δ	grain boundary thickness
$\delta(x)$	Dirac delta function
ϵ	surface energy
κ_{bf}	reaction constant at the interface
κ_{eff}	effective interfacial reaction barrier
κ_{mb}	reaction constant at the interface
λ	mean free path of electrons for inelastic scattering
λ_e	electron wavelength
μ	attenuation coefficient
$\frac{\mu}{\rho}$	mass-attenuation coefficient
ν	substitution variable
ξ	arbitrary, but fixed position
π	Pi
ρ	density
$\sigma_A(E_0, \beta, \Delta_A)$	partial scattering cross section for core-shell excitation of element A
$\sigma_B(E_0, \beta, \Delta_B)$	partial scattering cross section for core-shell excitation of element B
$\chi(u)$	phase distortion function
Ω	atomic volume

Introduction

Ceria and ceria-based materials are used as technical ceramics in different technological fields due to their favorable material properties. Examples are the use as catalyst or carrier for metallic catalyst particles due to their catalytic interactions with small molecules (H, CO, O, NO) or as Mixed Ionic-Electronic Conductor (MIEC) in oxygen sensors. CeO_2 is also investigated for application in electrochromic thin-film applications, medicine, as inert matrix fuel in reactors, as well as in Solid Oxide Fuel Cells (SOFCs).

Doping ceria with different Rare Earth (RE) atoms, especially Gd and Sm, strongly influences the oxygen-ion conductivity. This allows to control the (generally high) oxygen-ion conductivity of GDC and SDC. The high oxygen-ion conductivity facilitates the application of GDC and SDC in SOFCs as anode, electrolyte, and as diffusion barrier between the commonly used Y_2O_3-doped ZrO_2 (YDZ) electrolyte and Co-containing cathode layers.

Despite the high application potential of GDC and SDC, some basic materials properties are not well known. Oxygen-ion conductivities have been in detail studied in these materials, but few data are available on cation interdiffusion. Cation-interdiffusion coefficients and activation enthalpies for interdiffusion were up to now only derived from grain-growth experiments which can be strongly influenced by the formation of grain boundary phases or impurity segregation. The determined activation enthalpies range from 0.143 eV/atom to 9 eV/atom dependent on the used model. These findings motivate cation-interdiffusion studies on the basis of Gd_2O_3/CeO_2 and Sm_2O_3/CeO_2 diffusion couples which were performed for the first time in this work. The diffusion-couple geometry yields a well-established solution of the diffusion equation and allows straightforward evaluation of interdiffusion profiles.

Another aspect concerns the phase diagrams of GDC and SDC which are not well known for application-relevant temperatures below 1200 °C. $Gd_xCe_{1-x}O_{2-x/2}$ and $Sm_xCe_{1-x}O_{2-x/2}$ occur in different crystalline structures across the complete concentration range. GDC and SDC occur in the cubic fluorite phase at low and intermediate RE concentrations. The cubic bixbyite phase follows with increasing RE concentrations. Depending on the temperature the bixbyite structure or the monoclinic structure is stable at high RE concentrations. Possible miscibility gaps and a metastable cubic phase are still under debate, even at temperatures above 1200 °C. Hence, the investigation of the phase formation at lower application-relevant temperatures is interesting from a basic-science and application-relevant point of view.

Electron microscopy is a viable tool to study phase evolution and interdiffusion processes on the nanoscale. TEM allows to determine the local crystalline structure in the samples by high-resolution imaging and electron diffraction. Using STEM combined with analytical techniques, different phases in the samples can be characterized in detail. The nm-scale resolution of the analytical measurements additionally enables the quantitative determination of interdiffusion coefficients from interdiffusion profiles obtained from diffusion couples.

The present work is divided into four main parts. In Chapter 1 some basic material properties of GDC and SDC are presented. Then the different crystal structures and phase stability in the $Gd_xCe_{1-x}O_{2-x/2}$ and $Sm_xCe_{1-x}O_{2-x/2}$ systems are reviewed in detail. This is followed by an introduction to binary diffusion couples and temperature dependence of diffusion coefficients. Chapter 2 gives an overview of the electron microscopical techniques used in this work and presents the employed instrumentation. Chapter 3 contains the results on the $Gd_xCe_{1-x}O_{2-x/2}$ system which was studied in the temperature range from 986 °C to 1270 °C. The results of microstructural characterization and the measured concentration profiles for the $Gd_xCe_{1-x}O_{2-x/2}$ system allow conclusions on the phase evolution and facilitates the determination of interdiffusion coefficients. Analogous results for the more complex $Sm_xCe_{1-x}O_{2-x/2}$ system are presented in Chapter 4. This system was investigated in the temperature range between 987 °C to 1266 °C.

1 Fundamentals

In this chapter the application relevance of GDC and SDC as technical ceramic [1] is illustrated for the use in SOFCs, because Gd and Sm dopant cations strongly influence the oxygen-ion conductivity. The reader is referred to the books by Heinzel [2], Singhal [3], and Holtappels [4] for an introduction to SOFCs. Other applications which rely mainly on different material properties are not considered in this work (catalyst or carrier for metallic catalyst particles [5, 6], MIEC in oxygen sensors [7–9], electrochromic thin-film applications [10, 11], medicine [12], inert matrix fuel in reactors [13–15]). Then the phases in GDC and SDC are reviewed exhaustively. Special emphasis is placed on the stability ranges and crystal structures, where the results of previous studies differ significantly.

The basic principles used to characterize interdiffusion processes are introduced after the material characteristics with emphasis on diffusion couples. Furthermore, the Arrhenius type temperature dependence of diffusion processes is presented. Then grain growth models utilized to study activation enthalpies for cation diffusion are presented including results acquired by different groups.

Short Remarks Concerning Material Nomenclature

The material nomenclature in literature is often based on traditional terms. For example ceria is often used for CeO_2 but this name does not comply with the recommendation nomenclature by the International Union of Pure and Applied Chemistry (IUPAC). Additionally vague definitions are employed, e.g., GDC referring to $Gd_xCe_{1-x}O_{2-x/2}$ with a specific $x \approx 0.1$. These terms were used in the introduction to allow the reader already familiar with the broader research topic a quick overview in the way the reader is used to. To ensure clarity of the text, the meaning of the presented acronyms in literature is presented here and imprecise usage of terms in the following text is avoided.

Gadolinia and samaria are obsolete terms for the elemental formulae Gd_2O_3 and Sm_2O_3, which are mainly used when another material is doped or mixed with these oxides. Ceria is still in more widespread usage for the chemical formula CeO_2. Another expression found in literature is sesquioxide, which means the combination of a RE cation with oxidation state 3+ and an oxygen anion with oxidation state 2+, i.e., samarium(III) oxide.

Usually, GDC and SDC are used to refer to "gadolinium-doped ceria" and "samarium-doped ceria". The amount of dopant is then generally specified and in the majority of cases ranges between 5 at% and 30 at% of dopant. Instead of YDZ, the abbreviation Yttria-stabilized Zirconia (YSZ) is often used indicating the specific Y_2O_3-content of 8.5 mole%which was assumed to be sufficient for stabilization of the cubic phase.

In this work the somewhat imprecise nomenclature commonly used is avoided. Consequently, all compounds are named according to the recommendations of IUPAC in the following [16]. The formulae are given here in full rigour including oxidation states. However, in the following text the naming of compounds is given without oxidation states in accordance to IUPAC nomenclature for ease of reading. This means that

- for cerium(IV) oxide (formula $Ce^{IV}O_2^{II}$) in the following text the name cerium oxide and the formula CeO_2 are used and the term ceria is omitted.

- for gadolinium(III) oxide (formula $Gd_2^{III}O_3^{II}$) in the following text the name gadolinium oxide and the formula Gd_2O_3 are used and the term gadolinia is omitted.

- for samarium(III) oxide (formula $Sm_2^{III}O_3^{II}$) in the following text the name samarium oxide and the formula Sm_2O_3 are used and the term samaria is omitted.

- the abbreviation GDC in the following text is used to describe Gd_2O_3-doped CeO_2 with the chemical formula $Gd_xCe_{1-x}O_{2-x/2}$ and refers to the complete composition range $0 < x < 1$.

- the abbreviation SDC in the following text is used to describe Sm_2O_3-doped CeO_2 with the chemical formula $Sm_xCe_{1-x}O_{2-x/2}$ and refers to the complete composition range $0 < x < 1$.

- the abbreviation YSZ is not used at all in the following text. The correct YDZ is used to describe Y_2O_3-doped ZrO_2 instead.

The abbreviation Rare Earth (RE) is used for Sc, Y, and the lanthanoids (La, Ce, Pr, Nd, Pm, Sm, Eu, Gd, Tb, Dy, Ho, Er, Tm, Yb, and Lu) as approved by IUPAC [16].

1.1 Application of Gd$_2$O$_3$-doped CeO$_2$ (GDC) and Sm$_2$O$_3$-doped CeO$_2$ (SDC) in Electrochemical Devices

The study of GDC and SDC in this work is motivated by high application potential of these materials. They are characterized by a high oxygen-ion conductivity, which can be controlled by the dopant and the dopant concentration [17–24]. It was observed that GDC and SDC show the highest ionic conductivity compared to other RE dopants and YDZ at temperatures ranging from 500 °C to 1000 °C [19, 25]. This makes GDC and SDC viable candidates as oxygen conductors in SOFCs. At low oxygen partial pressures, doped CeO$_2$ is reduced and the ceramic becomes a MIEC [26]. This facilitates the application as anode material [27,28] and usually requires a noble metal catalyst for fuel reformation. However, an additional catalyst layer reduces the rate at which fuel can diffuse into the anode, and thereby decreases cell power density [29].

CeO$_2$-based ceramics were also investigated as an alternative electrolyte material to allow SOFC operation at lower temperatures (700 °C) compared to the commonly used YDZ [26, 30–32]. Another reason for this is related to the stability of the Y$_2$O$_3$-ZrO$_2$ material system which was studied extensively for decades [32–37]. Doping ZrO$_2$ with 8 mol% to 9 mol% of Y$_2$O$_3$ yields the highest oxygen conductivity in this system [38] and full stabilization of the cubic high-temperature even a room temperature was often assumed [39]. As a result YDZ has been established as one of the most commonly used electrolyte materials for SOFC applications for many years [32,40]. However, the ionic conductivity of YDZ decreases significantly within less than a few 1000 h [41–43]. Later studies then provided experimental evidence, that Spinodal decomposition of nanoscale precipitates is the reason for this performance degradation and the material is not fully stabilized at the mentioned dopant concentrations [44–47].

However, the usage of doped CeO$_2$ as electrolyte generates a new problem. The oxygen partial pressure is very low at the electrolyte/anode-interface, resulting in the reduction of the electrolyte. This reduction increases undesired electronic conductivity of the electrolyte [26]. Further lowering the operating temperature may be a solution to this problem, since the electronic conductivity is the negligible [32]. As a consequence the common Ni-YDZ anode has to be replaced and alternatives are not readily available [32].

Efforts were also undertaken to test doped CeO$_2$ as diffusion barrier between the common YDZ electrolyte and the cathode material due to its higher chemical stability regarding undesired electrolyte-cathode secondary phase formation. Experimenting with different material compositions and fuel cell configurations has not proven successfully until today [48–53].

The literature overview shows the wide range of applications of GDC and SDC. However there is a distinct lack of knowledge concerning the phase diagram and stable phases for the desired operating conditions. A late discovery of fundamental instabilities of the material system in the desired composition, as for YDZ, would be unfortunate. Therefore, investigations on the phase stability ranges in the $Gd_xCe_{1-x}O_{2-x/2}$ and $Sm_xCe_{1-x}O_{2-x/2}$ material systems are neccessary. A complete review of the knowledge on the phase diagrams is presented in the following Chapter 1.2.

1.2 Phases in Doped CeO$_2$ Systems

In this section the thermodynamic stability of doped ceria systems will be reviewed. At first, studies on the stability and crystal structure of pure cerium oxide are presented, since it is common to both investigated solid solutions. Then the properties of the solid solutions of CeO_2 with Gd_2O_3 and Sm_2O_3 are outlined. The following review papers are recommended as an introduction to the properties of binary RE oxides in general by Eyring [54] and Adachi and Imanaka [55] and on their phase diagrams by Zinkevich [56]. Bevan and Summerville summarize older studies on the mixing of rare earth oxides [57]. Crystallographic information is comprehensively compiled by Wyckoff [58] and Wells [59].

An overall feature of the whole series of the rare earth elements is the so-called lanthanide contraction, which refers to the continuous decrease in small steps of the ionic radii of the lanthanides and their corresponding oxides with increasing atomic number. As a result properties dependent on the ionic radii are similar due to the small variations of the ionic radii [60]. The similarity of the ionic radii of the RE oxide cations also leads to similar enthalpies of (phase) transition for the polymorphic structures of the RE oxides [61]. The resulting similar phase diagrams of $Gd_xCe_{1-x}O_{2-x/2}$ and $Sm_xCe_{1-x}O_{2-x/2}$ are introduced in the following. The relevant radii of six-fold coordinated Gd^{3+} and Sm^{3+} are 93.8 pm and 95.8 pm [62].

A Comment on the Classification of Crystal Structures

In the following text the crystal structure of the investigated materials is discussed in detail. The notation concerning the structure of RE oxides used in literature is inconsistent. The common usage of prototype structures since the first "Strukturbericht" by Ewald and Hermann in 1931 [63] and the standardized characterization by space group as recommended by the International Union of Crystallography (IUCr) [64] is not used consistently. It is mixed with the (arbitrary) structure designation introduced by Goldschmidt et al. [65] in 1925, who first investigated the RE oxides. Such notation is rejected by IUCr. In general using different types of notation lead to confusion. In this case the counter-intuitively named B-type structure is monoclinic, whereas the C-type corresponds to the bixbyite prototype structure. To enable coherent reading of the text, the original designation from 1925 is neglected. The prototype structures, space groups and crystal classes used for different structures are summarized here. The prototype structures are from Wyckoffs book [58].

The prototype for the cubic fluorite structure is the mineral fluorite CaF$_2$. It has the space group Fm$\bar{3}$m and is sometimes termed as F-type structure in literature.

The mineral bixbyite Fe$_x$Mn$_{1-x}$O$_{2-x/2}$ is the prototype for the cubic bixbyite structure and has variable iron and manganese content. It has the space group Ia$\bar{3}$ and Goldscmidt named it C-type structure in his work. In the case of Sm$_2$O$_3$ it is not clear if the space group is Ia$\bar{3}$ or I2$_1$3.

There is no prototype for the monoclinic structure of Sm$_2$O$_3$ and Gd$_2$O$_3$ and the two ceramic oxides are not used as prototype structures in literature themselves. The monoclinic structure has space group C2/m and also the term B-type structure is used in literature.

1.2.1 Pure CeO$_2$

Cerium oxide is the native oxide of the RE metal Ce with the chemical formula CeO$_2$ and is pale yellow in colour. First structural investigations were performed by Goldschmidt in 1923 [66]. X-Ray Diffraction (XRD) was used to determine the lattice parameter to 5.41 Å. Further analysis showed that CeO$_2$ has the fluorite structure. It is now well established that pure stoichiometric CeO$_2$ has the fluorite-type structure with space group Fm$\bar{3}$m over the whole temperature range from room temperature to the melting point at atmospheric pressure [24]. Cerium oxide tolerates a considerable reduction without phase change. However, CeO$_2$ is reduced for low oxygen partial pressures below 10^{-13} Pa in the temperature range between 600 °C and 1500 °C and forms a series of discrete stoichiometries Ce$_n$O$_{2n-2}$ with different crystal structures [67–70]. The reduction of CeO$_2$ under low O$_2$ partial pressure was investigated in situ by TEM by Yasunaga et al. [71]. Even for short times of electron irradiation in the temperature range of 23 °C to 200 °C, the growth of defect clusters due to the reduction was observed depending on electron energy (200 keV – 1000 keV) [71].

The ideal fluorite structure as used by Brauer & Gradinger [72] was assumed to simulate diffraction patterns of CeO$_2$ and the crystal structure data is given in Tab. 1.1.

Table 1.1: Lattice parameters and atomic positions of CeO_2, Gd_2O_3, Sm_2O_3, $Gd_{0.6}Ce_{0.4}O_{1.7}$, and $Sm_{0.5}Ce_{0.5}O_{1.75}$ including the reference from which the data for simulation was taken.

material	space group	lattice parameters [Å]	Positions of ions	Ref.
CeO_2 (fluorite)	$Fm\bar{3}m$	$c_1 = 5.411$	Ce $(0, 0, 0)$ O $(0.25, 0.25, 0.25)$	[72]
Gd_2O_3 (bixbyite)	$Ia\bar{3}$	$c_1 = 10.79$	Gd_1 $(-0.03144, 0, 0.25)$ Gd_2 $(0.25, 0.25, 0.25)$ O $(0.3915, 0.1524, 0.3809)$	[73]
$Gd_{0.6}Ce_{0.4}O_{1.7}$ (bixbyite)	$Ia\bar{3}$	$c_1 = 10.854$	Gd_1 $(0.25, 0.25, 0.25)$ Ce_1 $(0.25, 0.25, 0.25)$ Gd_2 $(-0.0188, 0, 0.25)$ Ce_2 $(-0.0188, 0, 0.25)$ O_1 $(0.388, 0.139, 0.376)$ O_2 $(0.401, 0.401, 0.401)$	[74]
Sm_2O_3 (bixbyite)	$Ia\bar{3}$	$c_1 = 10.934$	Sm $(-0.03144, 0, 0.25)$ Sm $(0.25, 0.25, 0.25)$ O $(0.3915, 0.1526, 0.3801)$	[73]
Sm_2O_3 (bixbyite)	$I2_13$	$c_1 = 10.93$	Sm_1 $(0.252, 0.252, 0.252)$ Sm_2 $(0.53, 0, 0.25)$ Sm_3 $(0.975, 0, 0.25)$ O_1 $(0.4, 0.145, 0.4)$ O_2 $(0.609, 0.841, 0.655)$	[75]
Sm_2O_3 (monoclinic)	$C2/m$	$c_1 = 14.17$ $c_2 = 3.63$ $c_3 = 8.84$ $\alpha = 99.96°$	Sm_1 $(0.1349, 0, 0.4905)$ Sm_2 $(0.1349, 0, 0.1380)$ Sm_3 $(0.4663, 0, 0.1881)$ O_1 $(0.128, 0, 0.286)$ O_2 $(0.324, 0, 0.027)$ O_3 $(0.299, 0, 0.374)$ O_4 $(0,469, 0, 0.344)$ O_5 $(0, 0, 0)$	[76]
$Sm_{0.5}Ce_{0.5}O_{1.75}$ (bixbyite)	$I2_13$	$c_1 = 10.81$	Sm_1 $(0.265, 0.265, 0.265)$ Ce_1 $(0.265, 0.265, 0.265)$ Sm_2 $(0.5, 0, 0.25)$ Ce_3 $(0.977, 0, 0.25)$ O_1 $(0.363, 0.13, 0.378)$ O_2 $(0.59, 0.859, 0.624)$	[77]

1.2.2 The Gd$_x$Ce$_{1-x}$O$_{2-x/2}$ System

In this chapter the crystalline structures of the Gd$_x$Ce$_{1-x}$O$_{2-x/2}$ system and its phase diagram are reviewed. At first the structural properties of pure Gd$_2$O$_3$ are presented and then data on the phase diagram of the complete material system is shown.

Pure Gd$_2$O$_3$

Gadolinium oxide is an inorganic compound with white colour of the RE metal Gd with the chemical fomula Gd$_2$O$_3$. The structure of Gd$_2$O$_3$ was studied extensively. Several different structures are reported for Gd$_2$O$_3$. It may exist in the cubic bixbyite structure and the monoclinic structure with space group C2/m under ambient conditions and the thermodynamic stability at low temperatures is still under debate [56]. The bixbyite and monoclinic structures transform into a hexagonal phase at sufficiently high temperature which converts to high-temperature cubic Gd$_2$O$_3$ before melting occurs at 2410 °C [78–80]. All structures were investigated in detail and here first the bixbyite and monoclinic structures are presented, followed by numerous observations on the temperature of the phase change from the bixbyite to the monoclinic structure.

Goldschmidt et al. [65] found the bixbyite structure for Gd$_2$O$_3$ at 600 °C and 750 °C and a monoclinic structure for 800 °C, 900 °C, and 1300 °C in 1925. The exact nature of the monoclinic structure was unclear and the space group Ia$\overline{3}$ with a lattice parameter of 10.79 Å was proposed for the bixbyite structure. Additional evaluation of this data according to Zachariasen [81,82] shows that the space group of the bixbyite structure is I2$_1$3. Pauling and Shappell [83] corrected this analysis and found that Ia$\overline{3}$ is the correct structure of bixbyite Gd$_2$O$_3$. Later measurements utilising XRD [84,85], perturbed angular correlation spectroscopy [73] and ab initio density functional theory [86] confirm the Ia$\overline{3}$ space group. The crystal structure is given in Tab. 1.1.

The monoclinic structure observed at higher temperature was investigated in more detail by Guentert and Mozzi [87]. They determined to it to be monoclinic with space group C2/m at temperatures above 1400 °C. This was confirmed later by Portnoi et al. [88]. However, here the transformation to the monoclinic phase already began between 1000 °C and 1300 °C. XRD performed by Curtis & Johnson [89] yielded the disappearance of the cubic structure at temperatures above 1300 °C with no structural analysis of the high-temperature phase. Another possibly trigonal structure mentioned by Ploetz et al. [90] was rejected due to possible influence from impurities. The stability temperatures for the different Gd$_2$O$_3$ structures are given in Fig. 1.1 (Gd-concentration of

100 mole%) along with the transition temperature range for the transformation from the bixbyite into the monoclinic structure.

This phase transition was studied especially regarding thermal equilibrium by several studies. Roth and Schneider [91] found a temperature of about 1250 °C and stated that the transformation from bixbyite structure to monoclinic structure is not reversible. Additional studies by Shafer and Roy [92], Warshaw and Roy [93], and Brauer and Müller [94] give different temperatures, shown in Fig. 1.1. All studies found a bixbyite→monoclinic transformation whereas the reversal was not found each time. If a monoclinic→bixbyite transition was found, the experimental conditions provided an alternative transformation path involving water, high pressure, or special starting material. This may indicate that the bixbyite structure is metastable. Systematical studies on the influence of pressure on the phase diagram suggest that high pressure stabilises the monoclinic phase [95, 96]. Later summarising work by Zinkevich [56] gives a bixbyite to monoclinic transformation temperature of 1152 °C. However, it is not elaborated how this value is derived from the large range of temperatures given by previous studies. The wide temperature range with measurements of both, the bixbyite and monoclinic structure or the transformation of one into the other (Fig. 1.1, right-hand axis), clearly shows that the equilibrium phase formation is strongly affected by the slow kinetics of transitions between different crystalline phases. The measurements which show the monoclinic phase below 1100 °C are probably due to (high) material impurities from other REs for the earlier studies.

Diffraction pattern simulations of pure Gd_2O_3 were performed using data on atomic positions refined by Bartos et al. [73] and are given in Tab. 1.1.

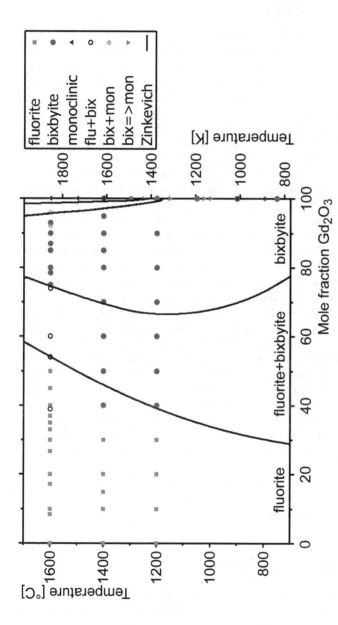

Figure 1.1: Phase diagram of the Gd$_x$Ce$_{1-x}$O$_{2-x/2}$ system with different measurements of different crystalline structures and single-phase and two-phase regions (green boxes for the fluorite structure (Fm$\bar{3}$m), red circles for the bixbyite structure (Ia$\bar{3}$), blue triangles for the monoclinic structure (C2/m), black empty circles for biphasic compositions with fluorite and bixbyite structure, cyan diamonds for biphasic compositions with bixbyite and monoclinic structure, and purple triangles for transition temperatures from the bixbyite to the monoclinic structure). Black lines delineate the phase boundaries predicted by Zinkevich [97].

The $Gd_xCe_{1-x}O_{2-x/2}$ Phase Diagram

The miscibility of Ce in Gd_2O_3 and vice versa was studied with varying results over the years. The studies cover the temperature range between 1200 °C and 1600 °C. The first study of Brauer and Gradinger [72] did not find a miscibility gap at 1400 °C and the change of the structure from the fluorite phase into the bixbyite phase occurs between 30 at% and 40 at%. Grover and Tyagi [98] confirmed the findings of Brauer and Gradinger at 1400 °C with a phase change between 40 at% and 50 at% utilising XRD. Investigating phase compositions at a lower temperature (1200 °C), Artini et al. [99] did not observe a two-phase region over the whole composition range using neutron diffraction. In correspondence to this work other studies found that it may be in fact not possible to state a composition at which the transformation from fluorite to bixbyite structure occurs. The reason for this is that the phase change occurs over the whole composition range and is not a sudden change, but an evolution in the sequence of clusters, domains, and precipitates with bixbyite structure in the range from 10 at% to 80 at% [100–102]. Another study proposed microdomains with a distorted pyrochlore structure, but this was not confirmed by any other study [103]. The sample preparation routes employed in these studies involved high pressure and may not be representative for thermal equilibrium.

Contradicting previous studies, Bevan et al. [104] found two miscibility gaps at 1600 °C. One is between the fluorite and bixbyite structure (composition range 54 at% – 74 at%) and the second between the bixbyite and monoclinic structure (composition range 92 at% – 100 at%). Similar results were also found by Zinkevich [97], who proposes a calculated $Gd_xCe_{1-x}O_{2-x/2}$ phase diagram. The calculation is based on his own measurements and previous measurements by Parks and Bevan. Several parameters had to be assumed for the calculation. The data from the studies is shown in Fig. 1.1 including the predicted phase diagram from Zinkevich. Slow transformation kinetics and the possibility of a metastable low-temperature phase prevent definite statements on miscibility gaps and stable equilibrium phases.

Structural data for diffraction pattern simulation of $Gd_{0.6}Ce_{0.4}O_{1.7}$ is shown in Tab. 1.1. The data is from an analysis of Grover et al. [74] who performed a Rietveld refinement of powder XRD data.

1.2.3 The $Sm_xCe_{1-x}O_{2-x/2}$ System

At first the literature on crystalline structures and phases of pure samarium oxide is discussed and the phase diagram of the $Sm_xCe_{1-x}O_{2-x/2}$ system is reviewed.

Pure Sm$_2$O$_3$

Samarium oxide is the oxide of samarium, a RE metal, with the chemical formula Sm$_2$O$_3$. The material has slightly yellow color in its stoichiometric form. The structure and equilibrium phases of of Sm$_2$O$_3$ were thoroughly investigated. Again the observations on structural properties are presented before a more detailed review is given on the stability ranges of the different phases. In general a behaviour similar to that of Gd$_2$O$_3$ is reported. The crystal structure of Sm$_2$O$_3$ under ambient conditions is under discussion [56]. The cubic bixbyite structure and the monoclinic structure are regarded as possible equilibrium phases. Increase in temperature stabilizes the monoclinic structure. Further increase of temperature results first in transformation to a hexagonal structure (which is the equilibrium phase for lower RE oxides) and then a transition to a different hexagonal high-temperature phase takes place at a temperatures above 2000 °C. A final phase change to the high-temperature cubic Sm$_2$O$_3$ occurs at about 2200 °C before the oxide melts at 2310 °C [78–80].

First investigations of Sm$_2$O$_3$ were carried out by Goldschmidt et al. [65] in 1925. The study identified the bixbyite structure at 620 °C, 630 °C, and 730 °C, the monoclinic structure at 900 °C, and a mixture of both at 735 °C. The spacegroup Ia$\bar{3}$ was proposed for the cubic bixbyite structure with a lattice parameter of 10.85 Å, whereas an identification of the monoclinic structure was not achieved. A detailed analysis of the data by Zachariasen [81, 82] claims that the correct space group of the bixbyite structure is I2$_1$3. A following study rejected this claim and states that Ia$\bar{3}$ as identified by Goldschmidt is the space group of bixbyite Sm$_2$O$_3$. Additional studies using different methods (XRD, angular correlation spectroscopy, ab-initio density functional theory, and Raman spectroscopy) supplied additional evidence that the bixbyite structure has the Ia$\bar{3}$ space group [73, 84–86, 105].

However, this evidence is contradicted by several other studies. Semiletov et al. [106] produced thin RE oxide films on NaCl substrates at room temperature and at 400 °C. The structure of the samarium sesquioxide films was then investigated by electron diffraction. Analysis of the data supports the claim by Zachariasen, that I2$_1$3 is the correct space group [75, 107]. It is noted that the I2$_1$3 space group is a sub-group of the Ia$\bar{3}$ space group of the bixbyite structure. In the Ia$\bar{3}$ structure each cation is surrounded by six oxygen ions at the corners of a highly distorted octahedron. Two types of octahedra exist in the Ia$\bar{3}$ structure which are different with respect to the cation-oxygen distances. The I2$_1$3 structure is almost identical to the Ia$\bar{3}$ structure. However, the cation-oxygen distances vary slightly within the octaedra. This results into an additional differentiation of the 48 oxygen atoms (per unit cell) in two classes of 24 oxygen ions each, giving rise to a symmetry reduction of the I2$_1$3 structure compared to the I$a\bar{3}$ structure. In the following the space group is stated in the text, if exact differentiation between the two identified space groups I$a\bar{3}$ and I$2_1 3$ is required.

The monoclinic structure occuring at higher temperatures was investigated by several authors. Samarium crystals fused in an oxyacetylene flame (3000 °C to 3500 °C) were produced and analysed by Douglass and Staritzky [108]. They state that the material is monoclinic. Additional XRD experiments with the same material by Cromer [76] yielded the C2/m space group. Later studies utilizing XRD of thin films and single crystals and powder neutron diffraction confirm these results [109–111]. The identification of the monoclinic structure as trigonal by Ploetz et al. [90] is rejected due to possible contaminations in the preparation route of the Sm_2O_3 (transformation from the bixbyite structure to the monoclinic structure above 750 °C). A phase transition to the low-temperature hexagonal structure (known from RE oxides with cation of lower atomic number (La, Ce, Pr, Nd) [56, 66]) at temperatures above about 1500 °C is also not considered since the oxides were in direct contact with water during the sintering process [92]. The structures determined at different temperatures are shown with the respective symbols in Fig 1.2.

Several studies focused on the phase transition from the low-temperature cubic bixbyite phase to the monoclinic phase at higher temperatures. Curtis and Johnson [89] performed XRD which shows the disappearance of a cubic structure at 1100 °C and higher temperatures. Portnoi et al. [88] and Roth and Schneider [91] found a non-reversible phase transformation from the cubic bixbyite structure to the monoclinic structure in temperature ranges between 900 °C and 1000 °C. A reversible phase transformation is suggested by Warshaw and Roy [93] at a temperature of 875 °C. Brauer and Müller [94] mainly investigate the bixbyite→monoclinic transformation and find that the minimum transformation temperature (here 850 °C) is strongly dependent on the preparation and sintering route of the starting material. In summary all studies found a bixbyite→monoclinic transition, whereas the reverse monoclinic→bixbyite transition is not reliably confirmed independent of preparation methods or other circumstantial proponents of the phase transformation, e.g. water or high pressure. A possible explanation is that the bixbyite structure is metastable. Systematic high pressure experiments suggest reversible monoclinic→bixbyite and monoclinic→hexagonal phase transformations [95, 112]. Sm_2O_3 thin films produced from samarium oxychloride (SmOCl) were investigated by Esquivel and coworkers [113]. XRD and electron diffraction show that the bixbyite→monoclinic transformation begins at a temperature of 800 °C. However, bixbyite Sm_2O_3 was still detected even after 1 month of annealing at 950 °C. The phase transformation in this study is influenced by the preparation route starting with tetragonal SmOCl (space group P4/mmm).

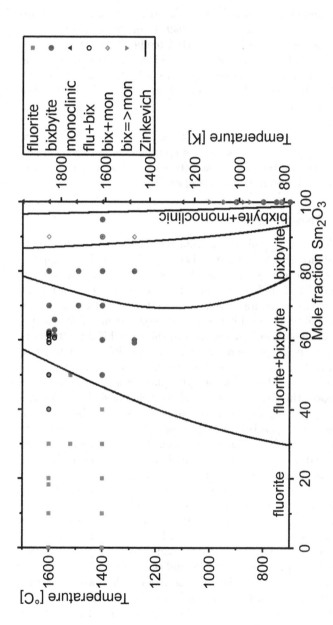

Figure 1.2: Phase diagram of the $Sm_xCe_{1-x}O_{2-x/2}$ system with different symbols indicating measurements of different crystalline structures and single-phase and two-phase regions (green boxes for the fluorite structure (Fm$\bar{3}$m), red circles for the bixbyite structure (Ia$\bar{3}$), blue triangles for the monoclinic structure (C2/m), black empty circles for biphasic compositions with fluorite and bixbyite structure, cyan diamonds for biphasic compositions with bixbyite and monoclinic structure, and purple triangles for transition temperatures from the bixbyite to the monoclinic structure). Black lines delineate the phase boundaries predicted by Zinkevich [97].

Zinkevich [56] reviewed some of the presented structure data to calculate phase diagrams and estimates a temperature of 949 °C for the bixbyite↔monoclinic phase transformation. Overall the reported temperature range of measurements with bixbyite and monoclinic structure and the bixbyite↔monoclinic phase transition (Fig. 1.2) is narrower for Sm_2O_3 than for Gd_2O_3 and the estimated value of 949 °C seems plausible. However, the transition kinetics between the different crystalline phases is still slow and has pronounced effects on equilibrium phase formation.

Structural data for simulation of diffraction patterns of pure Sm_2O_3 was taken from Bartos et al. [73] for the bixbyite structure with $Ia\bar{3}$ space group, from Zav'Yalova et al. [75] for the bixbyite structure with $I2_13$ space group, and from Cromer [76] for the monoclinic structure (see Tab. 1.1).

The $Sm_xCe_{1-x}O_{2-x/2}$ Phase Diagram

Research on the $Sm_xCe_{1-x}O_{2-x/2}$ phase diagram delivered inconsistent results regarding phase stability and miscibility. The data is limited to temperatures between 1400 °C and 1600 °C and several single measurements at 1280 °C. Brauer and Gradinger [72] found the fluorite structure for Sm-concentrations ≤ 40 at% and the bixbyite structure for Sm-concentrations ≥ 50 at% and state that there is no miscibility gap at 1400 °C. The measured monoclinic structure of pure Sm_2O_3 was not discussed and instead the cubic bixbyite structure was attributed to pure Sm_2O_3 in the discussion. The fluorite→bixbyite phase transition was confirmed by Mandal et al. [114] in the Sm-concentration range between 40 at% and 50 at%. The cubic bixbyite phase was not validated for high Sm-concentrations, though. A biphasic phase field consisting of bibyite and monoclinic phases is suggested at about 95 at%. A study by Bevan et al. [104] predicts miscibility gaps between the fluorite and bixbyite phases and the bixbyite and monoclinic phases. Later measurements gave evidence for two miscibility gaps at 1600 °C [115]. The first miscibility gap between the fluorite and bixbyite phases lies at Sm-concentrations of about 60 at%. An exact determination of phase boundaries was not possible. The second miscibility gap between the bixbyite and monoclinic phases occurs at Sm-concentrations ≥ 90 at% and was shown for 1280 °C and 1600 °C. Zinkevich [97] calculates a Sm_2O_3 phase diagram based on data of Parks and Bevans, own measurements, and several assumed parameters, which predicts wide miscibility gaps between the single phases. The complete data from all studies and the calculated phase boundaries are shown in Fig. 1.2. The possible metastable low-temperature bixbyite phase and the slow transformation kinetics again complicate explicit statements on equilibrium phases and miscibility gaps in a similar way as for the $Gd_xCe_{1-x}O_{2-x/2}$ system (see Chapter 1.2.2).

Structural data for diffraction pattern simulation is not readily available for $Sm_xCe_{1-x}O_{2-x/2}$ solid solutions with high Sm content and space groups $Ia\bar{3}$ and $I2_13$. To calculate diffraction patterns for comparison with experimental diffraction patterns, structural data of pure Sm_2O_3 was used for diffraction pattern simulations for both $Ia\bar{3}$ and $I2_13$ space groups.

It is neccessary to ensure that the introduction of Ce^{3+} ions does not change diffraction pattern simulations significantly. This was achieved by using structural data of the $Gd_xCe_{1-x}O_{2-x/2}$ system. A structural model of $Gd_{0.5}Ce_{0.5}O_{1.75}$ with space group $I2_13$ by Ye et al. [77] was adapted to $Sm_{0.5}Ce_{0.5}O_{1.75}$ by exchanging the Gd ions with Sm ions. The positions of the reflections in the simulated diffraction patterns of pure Sm_2O_3 and $Sm_{0.5}Ce_{0.5}O_{1.75}$ with bixbyite structure of space group $I2_13$ are equal. The data of $Sm_{0.5}Ce_{0.5}O_{1.75}$ is included in Tab. 1.1. The different number of cations results from the choice which Sm ions are substituted by Ce ions and the slightly different ion positions come from different results of the structural refinement.

The same result for the reflection positions as above holds for pure Sm_2O_3 and $Sm_{0.6}Ce_{0.4}O_{1.7}$ with bixbyite structure of space group $Ia\bar{3}$. Here again the data from Grover et al. [74] was used which was also used for the simulation of diffraction patterns of $Gd_{0.6}Ce_{0.4}O_{1.7}$. Thus $Sm_{0.6}Ce_{0.4}O_{1.7}$ has the same configuration as $Gd_{0.6}Ce_{0.4}O_{1.7}$ and is found in the respective entry in Tab. 1.1.

1.3 Binary Diffusion Couples

Binary diffusion couple are well suited to investigate phase formation as a function of composition. Additionally they allow the determination of cation interdiffusion coefficients. The neccessary concepts to discuss the experimental results of this work are introduced in the following chapter. This information is needed to extend the knowledge on phase diagrams.

1.3.1 Solution of the Diffusion Equation

A binary diffusion couple is considered as two semi-infinite bars differing in composition (e.g., two different metals) which are joined end to end at the plane $x = 0$. A mathematical description of the concentration of a diffusing atomic species C in dependence of position x and duration of the diffusion process t can be derived from Fick's laws and are given in textbooks on the subject, e.g. [116, 117]. A short introduction to the diffusion equation and the deduction of the diffusion couple solution is given here. The original work of Fick appeared in 1855 and Fick's laws are a purely phenomenological continuum description [118]. Fick's first law is

$$\vec{J} = -D(C)\nabla C. \tag{1.1}$$

It describes diffusion of an atomic species in isotropic media as a linear response between the concentration gradient ∇C and the atom or diffusion flux $\vec{J}$. The proportionality is described by the diffusion coefficient $D(C)$ in units $\mathrm{m^2 s^{-1}}$. The concentration gradient field ∇C is described by the nabla operator acting on the scalar concentration field $C(x, y, z, t)$. x, y, and z are position coordinates in 3-dimensional space. The concentration-gradient vector points in the direction, for which the concentration gradient is maximal. The concentration-gradient vector points in the opposite direction of the diffusion flux for isotropic media. Anisotropic media, a diffusion coefficient D(C) dependent from the concentration C, or chemical reactions of the diffusion atoms may cause deviations to Eq. 1.1.

Considering a diffusing atom species for an arbitrary test volume with varying diffusion fluxes J_i entering and leaving the volume, a difference in number of the atoms will occur if the fluxes do not balance (of course under the assumption that the number of atoms is conserved). The difference in inflow and outflow then results in a material accumulation (or loss) rate. Formulating this mathematically leads to a continuity equation

$$-\nabla \vec{J} = \frac{\partial C}{\partial t}. \tag{1.2}$$

Combining Eqs. 1.1 and 1.2 results in the second-order partial differential diffusion equation (or Fick's second law)

$$\frac{\partial C}{\partial t} = \nabla(D(C)\nabla C). \tag{1.3}$$

The equation is non-linear for concentration-dependent diffusion coefficients $D(C)$. An analytical solution is usually not possible, if the concentration dependence of $D(C)$ is random. Eq. 1.3 simplifies to

$$\frac{\partial C}{\partial t} = D\Delta C \tag{1.4}$$

for concentration independent D with the Laplace operator Δ. This equation can be further simplified by considering the geometrical arrangement of the experimental situation in this study. A thin film was deposited on substrate material with a grain size large enough (1 μm to 8 μm) to investigate diffusion far from grain boundaries. This allows to restrict the diffusion equation to one dimension since all positions at the interface are equivalent. Eq. 1.4 then leads to

$$\frac{\partial C}{\partial t} = D \frac{\partial^2 C}{\partial^2 x^2}.$$ (1.5)

Different initial conditions allow different solutions of Eq. 1.5. It is noted that D describes interdiffusion of the species in the diffusion couple which has to be distinguished from diffusion of the individual species. To emphasize this, in the following the interdiffusion coefficient $\tilde{D}$ is used. For a binary diffusion couple as described above the initial conditions are given by

$$C = \begin{cases} C_0, & \text{for} \quad x < 0 \\ 0 & \text{for} \quad x > 0. \end{cases}$$

To solve the diffusion equation for this initial distribution, the problem is solved at first for a slightly more convenient geometry, the so-called thin-film solution. Then the linearity of the concentration-independent diffusion equation is exploited which allows the derivation of new solutions by the principles of superposition. The starting condition ($t = 0$) of the thin-film solution is defined as

$$C(x,0) = N\delta(x).$$ (1.6)

which is also called instantaneous planar source. N denotes the number of diffusing atoms per unit area and $\delta(x)$ the Dirac delta function. The diffusing atoms are all placed at the plane $x = 0$ and allowed to spread for $t > 0$ into two material bodies occupying the half-spaces $-\infty < x < 0$ and $0 < x < \infty$. The solution of Eq. 1.5 with an initial concentration as defined in Eq. 1.6 and the atoms diffusing into both half-spaces is

$$C(x,t) = \frac{N}{2\sqrt{\pi Dt}} \exp\left(-\frac{x^2}{4\tilde{D}t}\right),$$ (1.7)

which is called Gaussian solution with the characteristic diffusion length $2\sqrt{\pi \tilde{D}t}$. From this solution, which restricts the experimental setup to a very thin layer of diffusing atoms in the beginning, the solution for a binary diffusion couple can be obtained. To this purpose the initial conditions as defined in Eq. 1.3.1 are interpreted as continuous distribution of instantaneous, planar sources of infinitesimal strength $dN = C_0 d\xi$ for $x < 0$. One unit length of the left-hand side contains $N = C_0 \cdot 1$ diffusing atoms per unit area and the right-hand bar contains no diffusant. The integral of all the infinitesimal responses resulting from the distribution of instantaneous source released from positions $\xi < 0$ can be thought as the solution $C(x,t)$ for this problem. The total reponse is given by the superposition

$$C(x,t) = C_0 \int_{-\infty}^{0} \frac{\exp[-(x-\xi)^2/4\tilde{D}t]}{2\sqrt{\pi \tilde{D}t}} d\xi.$$

Rearrangement of the equation results in

$$C(x,t) = \frac{C_0}{2}\left(1 - \mathrm{erf}\left(\frac{x}{2\sqrt{\tilde{D}t}}\right)\right).$$

Here erf denotes the error function

$$\mathrm{erf}(z) \equiv \frac{2}{\sqrt{\pi}} \int_0^z e^{\nu^2} d\nu,$$

and z and ν are given by

$$z = \frac{x}{2\sqrt{\tilde{D}t}} \quad \text{and} \quad \nu = \frac{x-\xi}{2\sqrt{\tilde{D}t}}.$$

Additionally definition of the complementary errorfunction as

$$\mathrm{erfc}(z) = 1 - \mathrm{erf}(z),$$

allows to write the binary diffusion couple solution in the common form

$$C(x,t) = \frac{C_0}{2}\mathrm{erfc}\left(\frac{x}{2\sqrt{\tilde{D}t}}\right). \tag{1.8}$$

The concentration profile is now described for one mobile species under the assumption of a concentration-independent interdiffusion coefficient $\tilde{D}$ and the exclusion of reactions during the diffusion process.

1.3.2 Binary Diffusion Couples in Multiphase Binary Systems

The solution of the diffusion equation for binary diffusion couples as described above is not sufficient to describe diffusion in the presence of more than one phase, e.g., if solidification or diffusional transformations occur or the starting materials are immiscible. At first a qualitative approach is chosen to understand and interpret diffusion profiles at interfaces of two materials which are not completely miscible mainly following the approach of Porter and Easterling [117]. The the case of interface-limited RL growth is presented afterwards.

Qualitative Description

Here the proposed phase diagram of Zinkevich [97] is considered as an example for two non-completely miscible materials as shown in Fig. 1.2. After annealing at 1600 °C, a diffusion couple made from Sm_2O_3 and CeO_2 will result in a layered structure containing three phases – the fluorite, the bixbyite, and the monoclinic phase. The phase distribution is shown as a hypothetical concentration profile in Fig. 1.3a). Note that here distance is plotted versus the Sm-concentration and that 100 at% Sm is on the right-hand side in contrast to the line scans presented later in this work (Chapter 4.3). The Sm-concentration varies from 0 at% to 53 at% in the fluorite phase, from 76 at% to 86 at% in the phase with bixbyite structure, and from 96 at% to 100 at% in the monoclinic phase. The just mentioned concentrations are the solubility limits of the phases at 1600 °C. The concentrations of 53 at% and 76 at% are seen to be the equilibrium concentrations of the fluorite and bixbyite phases in the fluorite+bixbyite two-phase field of the phase diagram. The fluorite and bixbyite phases are in local equilibrium across the fluorite/bixbyite interface. The same holds for the bixbyite/monoclinic interface at higher concentrations. These interfaces move as diffusion progresses. A hypothetical Sm-concentration profile is also shown in Fig. 1.3b) with concentration plotted versus distance to allow easy comparison with the experimental proiles acquired in this work. The concentration steps represent miscibility gaps (two-phase regions) in the phase diagram.

Reaction-Controlled Layer Growth

The influence of interfacial reaction barriers on the growth kinetics of layers, specifically in thin-film diffusion couples, is described inter alia by Gösele and Tu [119] and an introduction to the topic is also found in the textbook by Dybkov [120]. A combination of two processes determines the growth kinetics of the layer. The first process is the diffusion of matter across the layer, where the diffusion flux slows down with increasing layer thickness. The second process is the rearrangement of atoms at the interfaces required for the growth of the compound layer. This may involve a reaction barrier.

The effect of reaction-controlled RL formation is illustrated in the following. Here a layer of $Sm_\eta Ce_{1-\eta}O_{2-\eta/2}$ with bixbyite structure and thickness x_{bix} growing between two saturated phases is considered as an example. The monoclinic $Sm_\zeta Ce_{1-\zeta}O_{2-\zeta/2}$ and $Sm_\theta Ce_{1-\theta}O_{2-\theta/2}$ with fluorite structure have the saturation concentrations $C^{eq}_{\zeta(mon)}$ and $C^{eq}_{\theta(flu)}$. The interface positions to the bixbyite layer are marked by x_{mb} and x_{bf}. The subscripts $\zeta > \eta > \theta$ describe the composition of the compounds. The red line in Fig. 1.3 b) schematically shows the concentration profile of Sm in absence of interfacial reaction barriers. The

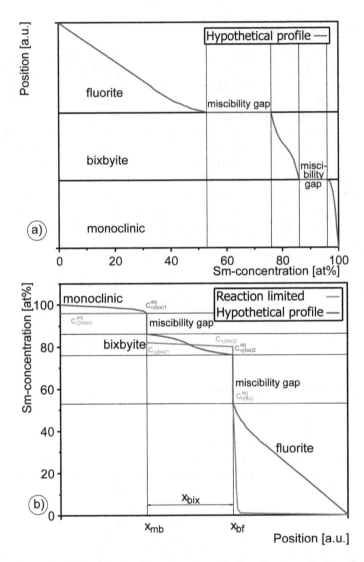

Figure 1.3: Possible diffusion layer structure for Sm_2O_3 deposited on CeO_2 after annealing at 1600 °C according to the proposed phase diagram from Zinkevich [97] as shown in Fig. 1.2. In a) distance is plotted against concentration to allow easy comparison with the phase diagram. In b) Sm-concentration is plotted against distance to allow easy comparison with the interdiffusion profiles presented in Chapter 3.3 and Chapter 4.3.

Sm-concentrations at the two interfaces correspond to the equilibrium concentrations $C^{eq}_{\eta(bix)1}$ and $C^{eq}_{\eta(bix)2}$ which give the maximum and minimum solubility of Sm in $Sm_\eta Ce_{1-\eta} O_{2-\eta/2}$ with bixbyite structure.

The green curve in Fig. 1.3 b) shows the Sm-concentration in the RL for reaction-controlled growth. The concentration gradient across the $Sm_\eta Ce_{1-\eta} O_{2-\eta/2}$ layer decreases in the presence of interfacial reaction barriers and the concentrations at the interfaces are then $C_{\eta(bix)1} < C^{eq}_{\eta(bix)1}$ and $C_{\eta(bix)2} > C^{eq}_{\eta(bix)2}$. The change of position of the two interfaces over time t is given by two equations under the assumption that the growth of the layer can be characterized by a constant interdiffusion coefficient $\tilde{D}$. The change of position for the interface beween the monoclinic and the bixbyite phase is

$$(C^{eq}_{\zeta(mon)} - C_{\eta(bix)1})\frac{dx_{mb}}{dt} = \tilde{D}\left(\frac{dC_\eta}{dx}\right)_{mb}, \tag{1.9}$$

where C_η is the Sm-concentration in the $Sm_\eta Ce_{1-\eta} O_{2-\eta/2}$ layer. And the change of position for the interface between the bixbyite and the fuorite phase is

$$(C_{\eta(bix)2} - C^{eq}_{\theta(flu)})\frac{dx_{bf}}{dt} = \tilde{D}\left(\frac{dC_\eta}{dx}\right)_{bf}. \tag{1.10}$$

From Eqs. 1.9 and 1.10 one can derive the change of the layer thickness x_{bix} introducing an effective interfacial reaction barrier κ_{eff} which is composed of the reaction constants at the interfaces κ_{mb} and κ_{bf}. The full derivation is not given here and it is noted, that one has to make the assumption $C_{\eta(bix)1} \approx C_{\eta(bix)2}$ to derive Eq. 1.11 which is given by

$$\frac{dx_{bix}}{dt} = K \cdot \kappa_{eff}(C^{eq}_{\eta(bix)1} - C^{eq}_{\eta(bix)2})(1 + \frac{x_{bix}\kappa_{eff}}{\tilde{D}})^{-1}, \tag{1.11}$$

where K is a constant depending on the composition of the involved phases. It follows from Eq. 1.11 that the described growth kinetics is different below and above the changeover thickness $x^*_{bix} = \tilde{D}/\kappa_{eff}$ with

$$x_{bix} \propto \begin{cases} t & \text{for} \quad x_{bix} \ll x^*_{bix} \\ t^{\frac{1}{2}} & \text{for} \quad x_{bix} \gg x^*_{bix}. \end{cases}$$

This means that at a sufficiently large thickness $x_{bix} \gg x^*_{bix}$ interface-controlled growth kinetics will always change to diffusion-controlled growth kinetics.

1.4 Temperature Dependence of Diffusion

1.4.1 The Arrhenius Relation

The Arrhenius relation was proposed by Savante Arrhenius in 1889. The Arrhenius equation gives the dependence of the rate constant k of a chemical reaction on the absolute temperature T in the following way

$$k = A \exp\left(-\frac{E_a}{k_B T}\right)$$

with the pre-exponential factor A, the activation energy E_a, and the Boltzmann constant k_B. This is an empirical relationship for thermally activated processes where the activation energy and the rate constant are experimentally determined. Subsequently, it was found that the Arrhenius equation is applicable to different temperature activated processes including diffusion [116]. In this case, the diffusion coefficient D obeys the Arrhenius formula

$$D = D_0 \exp\left(-\frac{\Delta H}{k_B T}\right) \tag{1.12}$$

with the frequency factor D_0 and the activation enthalpy ΔH. This is not always correct and has to be verified experimentally. Departures from simple Arrhenius behaviour may arise for different reasons like the mechanism of atomic migration, effects associated with impurities, and/or microstructural features such as grain-boundaries. Usually, an Arrhenius plot is used to illustrate this kind of temperature dependence. Here the natural logarithm of the diffusion coefficient is plotted versus the reciprocal temperature. This procedure results in a straight line (for Arrhenius type temperature dependence), which is immediately apparent by rewriting Eq. 1.12 to

$$\ln D = -\frac{\Delta H}{k_B}\frac{1}{T} + \ln D_0. \tag{1.13}$$

The logarithm of D obviously is linearly dependent on the reciprocal temperature. Now the y-intercept and the slope of the straight line can be determined from the experimental points using a linear regression. This in turn allows the calculation of D_0 and ΔH.

1.4.2 Grain Growth Experiments

Instead of diffusion-couple experiments, grain growth experiments were exclusively performed up to now to determine interdiffusion coefficients and activation enthalpies in the considered $Gd_x Ce_{1-x} O_{2-x/2}$ and $Sm_x Ce_{1-x} O_{2-x/2}$ systems.

Two models for the sintering processes were utilized. The main aspects of both formulations are presented shortly without giving the full mathematical derivations of the equations. Then the activation energies derived with these models are shown.

Isothermal Grain Growth

The kinetics of isothermal grain growth were deduced by Burke & Turnbull over 50 years ago [121,122]. Here grain growth is described as curvature process. The atoms diffuse along and across grain boundaries and the specific grain boundary area is reduced. This is mathematically expressed by a differential equation. The grain growth rate $\frac{dG}{dt}$ for parabolic grain growth is related to the mean radius of grain curvature which is proportional to the average grain size G by the following equation:

$$\frac{dG}{dt} = \frac{M\gamma}{G} \tag{1.14}$$

M denotes the grain boundary mobility and γ the grain boundary energy. Solving Eq. 1.14 leads to

$$G^2 - G_0^2 = k_p t \tag{1.15}$$

with the initial average grain size G_0. $k_p = 2M\gamma$ is a characteristic material constant. This can be generalized for non-parabolic grain growth which is often experimentally observed and was done by Yan et al. [123]. The differential equation is then

$$G^n - G_0^n = k_n t \tag{1.16}$$

where $k_n = nM\gamma G^{n-2}$ is a material constant. This can be further simplified with the assumption that $G_0^n << G^n$ can be neglected and that the constant k_n shows the Arrhenius type behaviour of a thermally activated process with

$$k_n = k_n^0 \exp\left(-\frac{Q}{k_B T}\right). \tag{1.17}$$

Q is the activation enthalpy of the grain growth process. Then Eq. 1.16 can be rearranged to

$$G^n = k_n^0 t \exp\left(-\frac{Q}{k_B T}\right). \tag{1.18}$$

This allows to determine the activation enthalpy Q using a regular Arrhenius plot as described in Chapter 1.4.1.

Combined-stage Sintering Model

The combined-stage sintering by Hansen et al. [124] model is based on the idea of Herring's scaling law [125] which states that particles remain geometrically identical during sintering besides a change in scale. The total atomic flux is calculated from Herrings general flux equation. This quantity is then used to calculate the macroscopic shrinkage of the material. This shrinkage is described by a differential equation containing intensive variables and microstructural parameters in the following way:

$$-\frac{dL}{Ldt} = \frac{\epsilon\Omega}{k_B T}\left(\frac{\delta D_b \Gamma_b}{G^4} + \frac{D_V \Gamma_V}{G^3}\right) \tag{1.19}$$

L is the sample length, ϵ the surface energy, Ω the atomic volume, and δ the grain boundary thickness. The diffusion coefficients D_b and D_V are related to grain boundary diffusion and volume diffusion, respectively. Γ_b and Γ_V describe, as collection of scaling factors, several features such as the ratio of the grain boundary area of the microstructure to the grain size or the diffusion distance at any given moment. Eq. 1.19 can be rewritten to a simpler form under the assuption that either grain boundary diffusion or volume diffusion is the dominant diffusion mechanism. The equation then reads

$$-\frac{dL}{Ldt} = \frac{\epsilon\Omega}{k_B T}\left(\frac{D_{bV}\Gamma}{G^n}\exp\left(-\frac{Q}{k_B T}\right)\right). \tag{1.20}$$

D_{bV}, Γ, and n correspond to the dominant diffusion mechanism with $D_{bV} = \delta D_b$ and $n = 4$ for grain boundary diffusion and $D_{bV} = D_V$ and $n = 3$ for volume diffusion. Taking the logarithm of Eq. 1.20 and slight rearrangement yield

$$\ln\left(-\frac{dL}{Ldt}T\right) = \ln\left(\frac{\epsilon\Omega D_{bV}\Gamma}{k_B G^n}\right) - \frac{1}{T}\frac{Q}{k_B}. \tag{1.21}$$

Now the activation enthalpy Q can be determined by plotting the term on the left-hand side versus $\frac{1}{T}$. This only works if Γ and G are independent of the material density.

Activation Enthalpies for Grain Growth of the $Gd_xCe_{1-x}O_{2-x/2}$ System

Two studies used the combined-stage sintering model where Eq. 1.21 was used to determine activation enthalpies for grain growth. Both studies investigated the same material composition ($Gd_{0.2}Ce_{0.8}O_{1.9}$). The first study by Jud et al. [126] reports activation enthalpies between 4.5 eV/atom and 9 eV/atom. However, this investigation cannot elucidate if the process is dominated by either grain boundary or cation volume diffusion. It is suggested that both processes contribute to the densification and that surface diffusion also plays a role. Additionally cation transport across grain boundaries may be impeded during grain growth. A subsequent study by Liang et al. [127] derives an activation enthalpy of 0.143 eV/atom. Since this was an unacceptable low value, another model by Young and Cutler [128] was used to fit the the the data. This results in an activation enthalpy of 5.4 eV/atom under the assumption that cation volume diffusion is the dominant diffusion mechanism. If grain boundary diffusion is assumed as predominant, a far too high activation enthalpy is found and the authors conclude that volume diffusion is the main driving force for grain growth.

A study by Gil et al. [129] assumed the kinetic grain growth to determine the activation enthalpy of $Gd_{0.1}Ce_{0.9}O_{1.95}$ which is expressed by Eq. 1.18. Their evaluation leads to an activation enthalpy of 5.36 eV/atom without indication of the dominant diffusion mechanism. A second study concerning thin films made of $Gd_{0.22}Ce_{0.78}O_{1.89}$ ascertained the activation enthalpy is 1.32 eV/atom employing isothermal growth studies [130]. This low activation energy was explained by self-limited grain growth where mainly grain boundary diffusion contributes to the grain coarsening.

Activation Enthalpies for Grain Growth of the $Sm_xCe_{1-x}O_{2-x/2}$ System

Only one investigation of grain growth is available for SDC. The grain growth of $Sm_{0.2}Ce_{0.8}O_{1.9}$ was analyzed by Okawa and Matsumoto [131]. Assuming kinetic grain growth the authors determine an activation enthalpy of 4.49 eV/atom for the material in the fluorite phase. The possible influence of grain boundary diffusion is not discussed. Additional problems are the difficulty to determine accurate starting grain sizes and deformation of the grains in an elongated shape during the sintering process.

Additional data for the grain growth in $Sm_xCe_{1-x}O_{2-x/2}$ co-doped with Co was acquired by Yan et al. [132]. However, the addition of Co strongly affected the sintering process and required a more complex model to describe the kinetics. Hence the data is not suited for comparison with the activation enthalpy derived from sintering of pure Sm_2O_3.

2 Experimental Techniques and Instrumentation

Methods and instruments used in this work are described in this chapter. Regarding TEM there are numerous techniques in addition to the basic functionality of imaging. The analytical techniques EDXS and EELS were used in combination with STEM to acquire interdiffusion profiles for evaluation. The two techniques are described in detail in the subsequent sections especially considering quantification of the experimental data. A basic overview of TEM is, for example, given in the books of Williams & Carter [133] and Reimer & Kohl [134]. SEM was used during the preparation process of the samples. The textbook by Reimer [135] explains SEM in detail.

2.1 Electron Microscopical Instrumentation

Microstructural characterization and HRTEM were conducted using two different transmission electron microscopes. The Philips CM200 FEG/ST microscope (now: FEI company, Hillsboro, Oregon, United States of America (USA)) was operated at 200 kV and is equipped with a Field Emission Gun (FEG) and a 4k×4k TemCam-F416 CMOS camera from TVIPS (Tietz Viedeo and Imaging Processing systems, Munich, Germany). The FEI Titan³ 80-300 microscope (FEI company) was operated at 300 kV and is equipped with an aberration corrector, for short C_s-corrector (CEOS – Corrected Electron Optical Systems GmbH, Heidelberg, Germany), in the imaging lens system. Single energy-dispersive X-ray spectra were collected using the NORAN Vantage system (Noran Instruments Inc., now: Thermo Fisher Scientific, Waltham, Massachusetts, USA) of the Philips CM200 FEG/ST microscope with a Ge X-ray detector and a probe diameter of about 2 nm. EDXS line scans were acquired utilizing the 30 mm EDAX Si(Li) X-ray detector with an ultra-thin window and an energy resolution of 136 eV (EDAX Inc., Mahwah, New Jersey, USA) in the FEI Titan³ 80-300 microscope. EELS was performed with the post-column Tridiem 865 HR Gatan Imaging Filter (GIF) (Gatan Inc., Pleasanton, California, USA) of the FEI Titan³ 80-300 microscope. EELS spectra can be with an energy resolution of 0.7 eV and a total channel count of 2048. The GIF was operated at a dispersion of 0.5 eV/channel for elemental quantification. SAED was performed on the Philips CM 200 FEG/ST microscope and the diffraction patterns were recorded on imaging plates from DITABIS (Digital Biomedical Imaging Systems AG, Pforzheim, Germany).

A LEO 1530 microscope (LEO Electron Microscopy Inc., now: Carl Zeiss NTS GmbH, Oberkochen, Germany) equipped with a FEG and a GEMINI® column was used for SEM.

Diffraction patterns were simulated with the software package Java Electron Microscopy Simulations (JEMS) version 3.3826U2009 by Stadelmann [136].

2.2 Transmission Electron Microscopy (TEM)

In TEM the sample is illuminated by a defocused electron beam which is transmitted through the sample and then used for imaging. The first transmission electron microscope of this type was built by Knoll and Ruska in 1932 [137]. It uses condensor lenses to produce the illuminating beam. A lens system below the sample is used to form the image with at least three lenses, i.e., an objective lens, an intermediate lens, and a projector lens. This arrangement can be used in two different modes – the imaging or the diffraction mode. They are selected by the excitation of the intermediate lens. It is either used to magnify the first intermediate image (imaging mode) or the diffraction pattern (diffraction mode) formed by the objective lens. The first intermediate image or the diffraction pattern is then further magnified by the projector lens onto an electron-sensitive scintillator and Charge-Coupled Device (CCD) camera which is used for detection.

The contrast in the final TEM image arises because of the scattering of the incident electrons by the specimen. Amplitude and phase of the electron wave can both be changed during its transit through the specimen and, thus, both can contribute to the image contrast. Hence, the fundamental distinction in TEM is between amplitude contrast and phase contrast. Both contrast types are normally involved in the image formation process. Usually, the imaging conditions are selected in a way that one contrast type dominates to be able to interpret the image accordingly.

Three different techniques used in this work are described in more detail. BFTEM and HRTEM are imaging techniques, whereas the diffraction mode is used for SAED. In-depth descriptions of these techniques are given in textbooks for TEM, e.g. [133, 134].

2.2.1 Bright-Field Transmission Electron Microscopy (BFTEM)

BFTEM is a method which is used to generate mass-thickness contrast or Bragg contrast. An aperture in the back focal plane is used to select only the un-diffracted electrons and exclude Bragg reflections from the image formation process. Mass-thickness contrast dominates the contrast in amorphous samples and crystalline samples only under kinematic excitation conditions. Bragg contrast dominates the contrast in crystalline materials. If mass-thickness contrast prevails, thicker sample areas or sample areas with higher mass (higher density or atomic number) scatter more electrons. Since scattered electrons do not contribute to the image, these sample areas appear darker than the surrounding area. If the imaging intensity is governed by Bragg contrast, the local intensity is determined by the excitation of Bragg reflections and the local sample thickness.

2.2.2 High-Resolution Transmission Electron Microscopy (HRTEM)

Phase contrast contributes significantly to image formation in HRTEM. The contrast arises due to coherent interference of contributions from at least two Bragg reflections. The interference pattern acquired by HRTEM reflects the periodicity of the crystal lattice (and is not a direct image of the crystal lattice). The resolution of transmission electron microscopes was limited by the spherical aberation of the lenses intil the late 1990s. The groundbreaking developement of a novel hexapole corrector by Rose and Haider for commercial microscopes made microscopes available with sub-Ångstrom resolution [138].

HRTEM requires the (crystalline) specimen to be oriented into a highly symmetrical (low-index) zone axis. Diffraction occurs and the atoms are arranged in columns parallel to the electron beam. A large objective aperture is used to select several Bragg beams which interfere and produce an interference pattern. This interference pattern is not a direct image of the atom positions and, therefore, cannot be readily interpreted.Intuitive interpretation in terms of atom positions is prevented by dynamic electron diffraction in the sample and the imaging process in the electron microscope. The influence of the microscope on the object wave function can be described by the contrast-transfer function. The contrast tranfer function includes the phase distortion function

$$\chi(u) = \frac{2\pi}{\lambda_e} \left(C_s \frac{\lambda_e^4 u^4}{4} + \Delta f \frac{\lambda_e^2 u^2}{2} \right).$$

It depends on defocus Δf, spatial frequency u, the electron wavelength λ_e, and the spherical aberration coefficient of the objective lens C_s. The spherical aberration can be corrected in aberration-corrected transmission electron microscopes.

HRTEM imaging requires thin samples because inelastic electron scattering has to be avoided. If atomic structure determination is intended, the experimental HRTEM images must be compared to simulations taking into account the local sample thickness and objective lens defocus. It is also possible to calculate the Fourier-transform from the image (or parts of it) which are called diffractograms. Diffractograms are used in this work for structure determination by comparison with simulated diffraction patterns. However, it must be taken into account, that additional Bragg reflections may appear in diffractograms due to dynamic electron diffraction and nonlinear image formation.

2.2.3 Selected Area Electron Diffraction (SAED)

The crystal structure can be determined from diffraction patterns using SAED. SAED allows to select a sample area from which a diffraction pattern is acquired by inserting an aperture in the first intermediate image plane. The sample is illuminated by a parallel beam and the microscope is operated in diffraction mode. The diffraction pattern is acquired by the CCD camera or imaging plates with a higher dynamical range. The diffraction pattern depends on the crystal structure and the orientation of the specimen.

The information contained in a SAED pattern is interpreted on the basis of the Bragg condition

$$2dsin\theta_b = n\lambda_e$$

which allows to extract the lattice plane distances. Size and symmetry of the unit cell determine the position of the Bragg reflections in the SAED pattern. The intensity of the reflections in the diffraction pattern depends on the structure factor and the lattice amplitude. Size and symmetry of the unit cell determines the position of the reflections in the diffraction pattern. The structure factor determines the occurence and intensity of reflections in kinematical diffraction which only applies if the intensity of the Bragg reflections is low in comparison to the undiffracted beam. The crystal structure is determined by orienting the specimen in different low-index zone axes, indexing the observed reflections, and comparing the result to simulated diffraction patterns.

Usually, dynamic electron diffraction has to be taken into consideration. Multiple scattering occurs with increasing sample thickness which leads to the appearance of kinematically forbidden reflections. In addition to dynamical effects, the number of inelastically scattered electrons increases with thickness. A reasonable method to reduce these effects is the selection of thin sample areas.

2.3 Scanning Transmission Electron Microscopy (STEM) and Beam Broadening

In STEM the sample is not illuminated by a broad beam as in conventional TEM. Instead a small probe is formed by focusing the electron beam to a small diameter and the sample is scanned sequentially. The signal of the scattered electrons transmitted through the sample is detected at every position. In a subsequent step the detected electron intensity from the scanned positions are assembled to an image of the sample where the local electron intensity determines the brightness of the corresponding pixel. The electrons are detected by different detectors depending on their scattering angle. The first scanning transmission electron microscope based on this functional principle was constructed by von Ardenne in 1938 [139, 140]. Usually, nowadays three different classes of detectors are distinguished. A bright-field detector is used for small scattering angles. Electrons scattered in larger angles are collected by annular dark-field detectors, where the largest scattering angles are covered by the High-Angle Annular Dark-Field (HAADF) detector. Due to the strong dependence of the total cross section for the scattering of electrons on the atomic number of the scatterer (here: the sample), imaging with strong material contrast is possible in the HAADF STEM mode where contribution of coherent Bragg diffraction can be neglected. Further detail on STEM imaging are given by Pennycook and Nellist [141].

In this work the contrast of STEM images was not eveluated directly. STEM imaging was mainly used to locate the interface between the thin film and the substrate. This allowed to acquire EDXS and EELS line scans across the interface using an electron probe with a diameter < 1 nm (spot size 6 in the FEI Titan^3 80-300 microscope [142]). However, beam broadening due to the electrons interacting with the specimen has to be taken into consideration in addition to the probe size itself which limits the spatial resolution of the composition analysis. The amount that the beam spreads on its way through the specimen can be approximated by the following formula [143] under the assumption that the electron undergoes only one elastic scattering event:

$$b = 8 \cdot 10^{-12} \frac{Z}{E_0} (N_V)^{\frac{1}{2}} t_S^{\frac{3}{2}} \tag{2.1}$$

Here b is the beam spread in nm, Z the atomic number, E_0 the primary electron energy in keV, N_V the number of atoms/m^3, and t_S the sample thick-

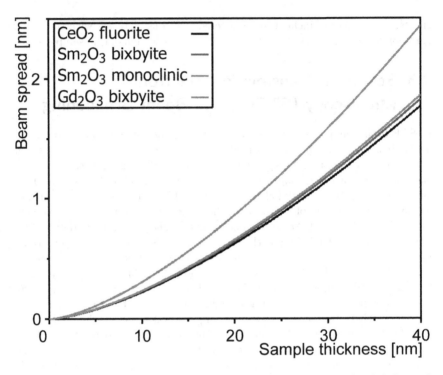

Figure 2.1: Beamspread calculated according to Eq. 2.1 for CeO_2, Gd_2O_3, and Sm_2O_3 with material parameters listed in Tab. 2.1.

ness. The mean atomic numbers $\overline{Z}$ of the different investigated materials were used to calculate the beam spread. The primary electron energy is 300 keV. The constant parameters $\overline{Z}$ and N_V of Eq. 2.1 are listed for CeO_2, Gd_2O_3, and Sm_2O_3 (with different crystalline structures) in Tab. 2.1.

A plot of the beam spread up to a sample thickness of 40 nm is shown in Fig. 2.1. The maximum beam spread is clearly expected for monoclinic Sm_2O_3. The beam spread of the mixed systems $Gd_xCe_{1-x}O_{2-x/2}$ and $Sm_xCe_{1-x}O_{2-x/2}$ with bixbyite structure correspond approximately to the beam spread of the pure materials (CeO_2, Gd_2O_3 and Sm_2O_3) with fluorite and bixbyite structure. A maximum local sample thickness of 38 nm is determined in Chapter 2.5.1 for all investigations which lead to quantitative evaluations of interdiffusion coefficients. This thickness yields a beam spread of 2.25 nm for monoclinic Sm_2O_3 and about 1.65 nm for the other structures.

2.4 Quantitative Energy-Dispersive X-ray Spectroscopy (EDXS)

The idea of using X-rays for elemental analysis in electron-beam instruments was first described by Hillier and Baker in 1944 [144]. The outstanding dissertation by Castaing [145] was devoted to the construction of the required instrumentation. In this work the Cliff-Lorimer method was used for quantification of the EDXS spectra [146]. This ratio technique developed by Cliff and Lorimer in 1975 relates the elemental concentrations of different elements and the element characteristic X-ray intensity by the following equation:

$$\frac{C_A}{C_B} = k_{A,B}\frac{I_A}{I_B} \tag{2.2}$$

Here C_A and C_B are the concentrations of element A and element B and I_A and I_B the peak intensities of the chosen characteristic X-ray line of element A and element B. Additionally the sum of the concentrations of the elements is assumed to constitute 100 % of the sample, meaning

$$C_A + C_B = 100\% \tag{2.3}$$

The Cliff-Lorimer factor $k_{A,B}$ is a sensitivity factor which depends on the characteristic X-ray lines considered for the investigation, the experimental setup, and the elements involved. The FEI TEM Imaging and Analysis (TIA) software suite (version 4.3 build 904) used for EDXS quantification supplies a database of $k_{A,B}$ factors for quantitative investigations. A consistency check of the available $k_{A,B}$ factors was made in this thesis which involves EELS as described in the respective Chapter 2.5.

The $L_\alpha-$ and $L_\beta-$ X-ray-emission lines of Ce, Gd and Sm, which were used for quantification, are tabulated in Tab. 2.2. The values of the X-ray energies are from Bearden [147]. The emission lines of Ce and Gd do not overlap, and the $L_{\alpha 1}$ and $L_{\alpha 2}$ lines of Ce and Sm are also clearly separated. However, there

Table 2.1: Constant parameters $\overline{Z}$, N_V, and crystal structures of different materials.

material	crystal structure	space group	$\overline{Z}$	N_V [atoms/m³]
CeO_2	fluorite	Fm$\overline{3}$m	30	$7.62078951379363 \cdot 10^{28}$
Sm_2O_3	bixbyite	Ia$\overline{3}$	34.4	$6.17746784048851 \cdot 10^{28}$
Sm_2O_3	monoclinic	C2/m	34.4	$1.10154090324473 \cdot 10^{29}$
Gd_2O_3	bixbyite	Ia$\overline{3}$	35.2	$6.12001911107205 \cdot 10^{28}$

Table 2.2: X-ray emission lines of Ce, Gd, and Sm used for EDXS quantification

X-ray energy [keV]	Ce	Gd	Sm
4.8230	$L_{\alpha 2}$		
4.8402	$L_{\alpha 1}$		
5.2622	$L_{\beta 1}$		
5.2765	$L_{\beta 4}$		
5.3651	$L_{\beta 3}$		
5.6090			$L_{\alpha 2}$
5.6134	$L_{\beta 2}$		
5.6361			$L_{\alpha 1}$
6.0250		$L_{\alpha 2}$	
6.0572		$L_{\alpha 1}$	
6.1960			$L_{\beta 4}$
6.2051			$L_{\beta 1}$
6.6871		$L_{\beta 4}$	
6.7132		$L_{\beta 1}$	

is an overlap between the Sm-L_α emission lines and the Ce-$L_{\beta 2}$ emission line. This can be corrected for because the known relative intensity of the Ce-$L_{\beta 2}$ is 21 % [148].

Absorption and secondary fluorescence of X-rays in the samples may affect quantification. Photon mass absorption was first investigated by Bouguer [149] and in more depth by Lambert [150]. The phenomenological Beer-Lambert law is shown here in the form which is used when discussing material-specific mass attenuation coefficients (for example by Hubbell [151]):

$$\frac{I_{att}}{I_{inc}} = \exp\left(\frac{\mu}{\rho}l\right) \tag{2.4}$$

The incident intensity of photons I_{inc} is attenuated to the intensity I_{att} in passing through a layer with mass-per-unit area or area density $l = \rho t_l$. t_l is the layer thickness and ρ the density of the layer material. The process of attenuation is dependent on the material- and energy-dependent attenuation coefficient μ. The mass attenuation coefficient is given by $\frac{\mu}{\rho}$. The mass attenuation coefficient is the weighted sum of the atomic constituents or homogeneous mixtures and compounds with

$$\frac{\mu}{\rho} = \sum_i w_i \left(\frac{\mu}{\rho}\right)_i, \tag{2.5}$$

where w_i is the fraction by weight of the ith atomic constituent [152]. Eq. 2.5 means that the compound with the maximum mass attenuation coefficient in the relevant X-ray energy range between 4.8 kV and 6.8 kV (as listed in Tab. 2.2) limits the thickness of the sample regarding absorption and secondary fluorescence. The database specialized on crystallography and XRD by Chantler [153, 154] provides the maximum value of $\left(\frac{\mu}{\rho}\right)_{Ce,max} = 504 \text{ cm}^2\text{g}^{-1}$ for CeO_2 in the given energy range. The database is also available online in convenient form [155]. The mass attenuation coefficients of Gd_2O_3 and Sm_2O_3 are lower in this energy range and, thus, also of the compound systems $Gd_xCe_{1-x}O_{2-x/2}$ and $Sm_xCe_{1-x}O_{2-x/2}$. $\left(\frac{\mu}{\rho}\right)_{Ce,max}$ allows to calculate the thickness for an absorption of 1 % of the X-ray intensity. This occurs at a TEM sample thickness of about 60 nm. The thickness of the investigated samples is estimated to be lower than 38 nm (c.f. Chapter 2.5). Hence, absorption can be neglected for quantitative EDXS analyses.

Quantitative composition analysis was performed with the TIA software. To remove the background from the EDXS spectra a 5th order polynomial was fitted to the signal in chosen energy windows between the X-ray lines. This polynomial was used to extrapolate the background signal under the characteristic X-ray peaks and to subtract the background from the signal. Gaussian curves were fitted standard-less to the characteristic peaks of the L_α- and L_β-lines of the background corrected data to obtain the intensity (counts) of the X-ray lines. This fit was reiterated until the fit parameters converged to the same values for subsequent iterations. The cation concentrations were evaluated from the determined intensities using the $k_{A,B}$ factors implemented in the TIA software. Since these factors yield compositions in weight percent, all determined concentrations were converted to atomic percent (at%).

The complete concentration data presented in this study refers to cation concentration normalized to 100 at% as given in Eq. 2.3. The concentration of O-atoms on the anion sublattice is not evaluated as quantification of light elements is prone to introduce large errors owing to the low X-ray yield of light elements. The systematic error of the cation concentrations stems from inaccuracies in peak fitting and background subtraction as well as absorption effects. Also the preparation process including Ar^+-ion milling may introduce measurement artifacts. In general the total systematic error of the cation concentrations is estimated to be ± 3 at% for the complete data shown in this work. The statistic error is assessed by multiple measurements of diffusion profiles at different sample positions and calculation of the standard deviation of the determined values. The total error is given together with the corresponding data in the following.

2.5 Electron Energy Loss Spectroscopy (EELS)

EELS is a technique which can be used for analysis of several different material properties in TEM. To this purpose the energy distribution of electrons transmitted through the sample is measured by a magnetic-prism spectrometer. These electrons may have undergone inelastic scattering (loosing energy). The energy-loss events yield detailed information about the chemical composition of the specimen, the sample thickness, and the electronic structure of the specimen atoms including, amongst others, the valence state, the dielectric response, and the free-electron density. First measurements of the kinetic energy of electron utilizing a magnetic spectrometer were performed with reflected electrons by Rudberg [156]. Ruthemann [157] later used electrons transmitted through a thin film to measure low-loss spectra. The first instrument to achieve elemental analysis was constructed by Hillier and Baker [144]. An introduction to EELS can be found in the book by Williams and Carter [133]. Methods for thickness determination and composition analysis are presented in detail by Egerton [158] and Brydson [159]. A fine overview on instrumentation and reference spectra is found in the textbook edited by Ahn [160].

The basic component in all EELS and energy-filtering systems is the magnetic prism. It produces the spatial separation of the electrons according to their energy. The magnetic spectrometer in modern transmission electron microscopes is implemented in two different ways – either as an in-column filter or as a post-column spectrometer. Both allow operation in spectroscopic mode with parallel acquisition of the complete energy-loss spectra (Parallel-Collection Electron Energy Loss Spectroscopy (PEELS)) or in imaging mode which allows the acquisition of energy-filtered images. An in-column filter is passed by all electrons during microscope operation, because it is incorporated in the microscope column in the optical path. Aberrations may be introduced in the imaging system by the lens properties of this kind of spectrometer and image artifacts will also be present for unfiltered applications. The main advantages of in-column filters are high count rates for energy-filtered imaging and the possibility of filtered images and diffraction patterns in all imaging planes of the microscope. Post-column spectrometers are attached as an additional piece of equipment after the viewing screen and/or CCD camera. No aberrations are created in the beam path before entry into the GIF. In this work a post-column spectrometer was used to acquire electron energy-loss spectra.

Two different parts of the spectra are considered for the evaluation of sample properties in general. The low-loss region up to an energy loss of about 50 eV is distinguished from the high-loss region with energy losses ranging from 50 eV to 2000 eV. The low-loss region including the ZLP and the plasmon peak contains electronic information from the more weakly bound conduction and valence-band electrons. Parameters like sample thickness or the dielectric permittivity and electronic properties like covalent bond excitations can be determined from the low-loss spectra. The high-loss region (also named core-loss region) shows the element-specific ionization edges which give informations on the chemical composition of the sample. The ionization edges exhibit steep sharp double peaks in certain elements which are called white lines. The sharp peaks arise because the core electrons are excited into well-defined empty states (and not a broad continuum) for the $L_{2,3}$ edges of the transition metals and the $M_{4,5}$ edges of the rare-earth elements. The peaks of the ionization edges are superimposed on a background that decays exponentially with energy loss.

In this study EELS investigations were performed to measure the sample thickness and elemental compositions. The sample thickness is relevant for EDXS and EELS quantification. The chemical composition along linescans across the Sm_2O_3/CeO_2 interfaces were performed as an additional consistency check of the compositions derived by EDXS. The techniques for quantitative analysis and thickness determination are presented in the following.

2.5.1 Thickness Determination

The influence of sample thickness on beam broadening and the spatial resolution of the interdiffusion profiles was estimated by measuring low-loss spectra at the investigated sample areas. Information on the local sample thickness is also important to ensure that multiple scattering that may have a strong influence on quantitative EELS is negligible. An incident electron may undergo multiple inelastic scattering events during its transmission through the sample. The probability of an inelastic scattering event increases with the sample thickness t_s. Assuming independent scattering, the electron intensity, integrated over energy loss, follows a Poisson distribution. The additional constraint that only one scattering event occurs (single scattering) allows to describe the fraction of electrons which remain unscattered by the following equation:

$$J_{ZLP} = J_{tot} \exp\left(-\frac{t_s}{\lambda}\right). \tag{2.6}$$

Here J_{ZLP} is the electron intensity of the ZLP and J_{tot} is the total electron intensity collected up to an energy loss of about 200 eV. The GIF was operated at a dispersion of 0.1 eV/channel with 2048 channels to disperse the electrons

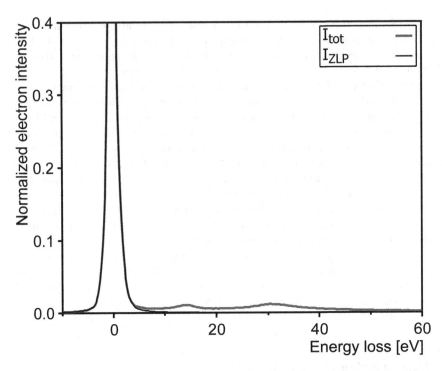

Figure 2.2: Low-loss spectrum from $Sm_xCe_{1-x}O_{2-x/2}$ (sample Sm-1175) with a relative thickness of 0.3 used for thickness determination.

of the high-intensity ZLP into a higher amount of channels compared to a dispersion of 0.5 eV/channel. This reduces the risk of damage to the spectrometer. The difference of J_{tot} to the intensity of all electrons which include energy losses over 200 eV can be neglected. λ is the mean free path of electrons for inelastic scattering and $\frac{t}{\lambda}$ is the relative thickness of the sample. A low-loss spectrum used to determine the relative thickness is shown in Fig. 2.2 and illustrates the proportion of J_{tot} to J_{ZLP}. J_{ZLP} is determined by assuming a symmetrical ZLP and fitting a Gaussian around the energy loss of 0 eV.

Eq. 2.6 can be rewritten in the form

$$\frac{t_s}{\lambda} = \ln\left(-\frac{J_{tot}}{J_{ZLP}}\right). \tag{2.7}$$

Eq. 2.7 was used to determine the local relative thickness of the samples. Regions were chosen where the relative sample thickness did not change significantly across the interface between CeO_2 and Gd_2O_3 or Sm_2O_3, apart from statistical fluctuations. All sample regions where interdiffusion was investigated have a relative thickness $\frac{t_s}{\lambda} < 0.3$. This reasonably small values allow to neglect multiple scattering. No further processing of core-loss spectra is required to correct thickness contributions, i.e. multiple scattering, to the measured core-loss spectra. If the mean free path for inelastic scattering λ is known, the actual sample thickness can be calculated from Eq. 2.7. Malis et al. [161] proposed parametrization to estimate λ in nm:

$$\lambda = \frac{106 \cdot F E_0}{E_m \ln\left(\frac{2\beta E_0}{E_m}\right)}. \tag{2.8}$$

λ is dependent on the relativistic factor F, the primary electron energy E_0 [keV], the average energy loss E_m [eV] , and the collection semi-angle β. F and E_m are calculated by the following equations

$$F = \frac{1 + \frac{E_0}{1022}}{(1 + \frac{E_0}{511})^2}, \tag{2.9}$$

$$E_m = 7.6 \text{ eV} \cdot Z_{eff}^{0.36}. \tag{2.10}$$

The effective atomic number Z_{eff} in Eq. 2.10 is determined by the atomic numbers Z_i of the elements constituting the sample by the following equation:

$$Z_{eff} = \frac{\sum_i a_i Z_i^{1.3}}{\sum_i a_i Z_i^{0.3}}. \tag{2.11}$$

The atomic numbers Z_i in Eq. 2.11 are weighed by their atomic fraction a_i. The relevant elements are Ce, Gd, Sm, and O with the atomic numbers 58, 62, 64, and 8. All additional parameters in Eqs. 2.8 to 2.11 are known. This allows to estimate the mean free paths of the pure compounds CeO_2 (105 nm), Gd_2O_3 (99 nm), and Sm_2O_3 (100 nm). Values of λ for $Gd_xCe_{1-x}O_{2-x/2}$ and $Sm_xCe_{1-x}O_{2-x/2}$ lie in between the mean free paths of the thin film materials with $\lambda_{Gd_2O_3} < \lambda_{Gd_xCe_{1-x}O_{2-x/2}} < \lambda_{CeO_2}$ and $\lambda_{Sm_2O_3} < \lambda_{Sm_xCe_{1-x}O_{2-x/2}} < \lambda_{CeO_2}$. Allowing an error margin of 20 % on the maximum mean free path λ_{CeO_2} of the investigated materials, the maximum local thickness of the investigated TEM samples can be estimated to 38 nm.

2.5.2 EELS Composition Analysis

Background Correction

As mentioned above the element-specific ionization edges are superimposed on the falling background of preceding ionization edges (if present) and the plasmon excitations at lower energy losses. Due to the complexity of these contributions, a model of the background from first principles is not available [158,159]. However, there are simple methods to remove the background from the signal. In energy-loss ranges sufficiently beyond preceding ionization edges the background signal J_{Back} can be estimated by a power law of the following form:

$$J_{Back} = B \cdot \Delta E^{-r}. \tag{2.12}$$

ΔE denotes the energy loss of the transmitted electrons, whereas B and r are fit parameters. Fitting Eq. 2.12 to the pre-edge background of the respective ionization edges, allows to correct the background intensity. The energy-loss ranges for background fitting are 785 eV to 860 eV for the Ce-$M_{4,5}$ and 970 eV to 1070 eV for the Sm-$M_{4,5}$ ionization edges. The signal curve, the fitted backgound curve, and the resulting corrected signal are shown in Fig. 2.3. Small fluctuations around an electron intensity of 0 of the pre-edge residual signal after background subtraction was used as criterion for a good background fit.

Composition Quantification

Similar to the characteristic X-ray emission peaks measured by EDXS (see Chapter 2.4), EELS allows to relate measured electron intensities of the element-specific ionization edges to the local chemical composition. In the case of EELS, the measured electron intensity of a selected ionization edge of an element A J_A depends on the probability for the excitation of a core-shell electron and the number of atoms of element A in the probed sample volume. This dependence can be described using a partial scattering cross section $\sigma_A(E_0, \beta, \Delta_A)$ by the following equation:

$$J_A(\beta, \Delta_A) = N_A \cdot J_{ZLP}(\beta) \cdot \sigma_A(E_0, \beta, \Delta_A). \tag{2.13}$$

N_A is the number of atoms per unit area, J_{ZLP} the intensity of the ZLP and the partial scattering cross section σ_A depends on the primary electron energy E_0, the collection semi-angle β, and the integration window over a limited energy range Δ_A. Eq. 2.13 includes the assumption that only single scattering contributes to the core-loss edges (or plural scattering is removed). A concentration ratio of elements A and B is more commonly determined. Thus the number of atoms per unit area N_A and N_B are replaced by the concentrations of the respective elements C_A and C_B. Dividing the element-specific measured intensities J_A and J_B from Eq. 2.13 and resolving for the concentrations leads to

$$\frac{C_A}{C_B} = \frac{\sigma_B(E_0, \beta, \Delta_B)}{\sigma_A(E_0, \beta, \Delta_A)} \cdot \frac{J_A(\beta, \Delta_A)}{J_B(\beta, \Delta_B)}. \tag{2.14}$$

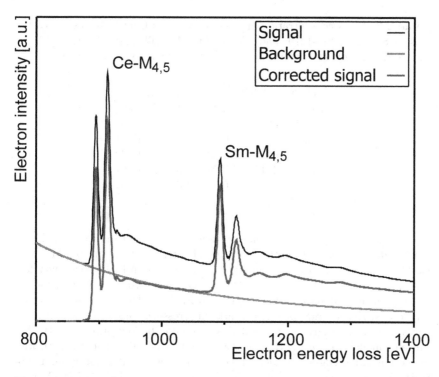

Figure 2.3: Exemplary electron energy-loss spectrum obtained from the RL1 of the Sm-1170 sample.

Here the intensities J_A and J_B may arise from different core-shell excitations of the elements A and B. The energy-loss windows Δ_A and Δ_B which are used for integrating the net signal of a specific ionization edge commonly start at the onset of an ionization edge. Comparing Eq. 2.14 to the Cliff-Lorimer equation (Eq. 2.2) used for EDXS quantification in Chapter 2.4, one can determine a sensitivity factor s with

$$s = \frac{\sigma_B(E_0, \beta, \Delta_B)}{\sigma_A(E_0, \beta, \Delta_A)} \tag{2.15}$$

analogous to the Cliff-Lorimer factor $k_{A,B}$. However, it is important to use the same set of experimental parameters E_0, β, Δ_A, Δ_B and exposure time of the spectrometer t_e for all EELS measurements to prevent the introduction of measurement artifacts into the quantification process and this was thoroughly taken care of.

Table 2.3: Mean concentration and intensity ratios from linescans across the RLs of Sm-1170, Sm-1219, and RVs of Sm-1175-20h (see Chapter 4.1.3 for sample denotations) to determine the sensitivity factor $s_{Sm,Ce}$.

sample and region	$\overline{\dfrac{C_{Ce}}{C_{Sm}}}$	$\overline{\dfrac{J_{Ce}}{J_{Sm}}}$	$\dfrac{1}{s_{Sm,Ce}}$
Sm-1170 RL1	0.9996	1.68	1.681
Sm-1170 RL2	0.227	0.375	1.652
Sm-1219 RL1	0.98	1.657	1.691
Sm-1219 RL2	0.239	0.414	1.732
Sm-1175-20h RV	0.248	0.412	1.661
Sm-1070	0.235	0.391	1.664

In this work Sm- and Ce-concentrations were derived from the net intensities (after background subtraction, see Chapter 2.5.2) of the Sm-$M_{4,5}$ and Ce-$M_{4,5}$ ionization edges J_{Sm} and J_{Ce} on the basis of an EELS sensitivity factor $s_{Sm,Ce}$ according to Eq. 2.15. The net EELS intensity curve used for the Ce-concentration determination is shown in Fig. 2.3. This procedure was chosen because the accuracy of partial scattering cross-sections for these ionization edges depends strongly on the model used to describe ionization, e.g., the hydrogenic model [162] or the Hartree-Slater model [163–165]. The net intensities of the ionization edges were determined by integration after background subtraction as described in Chapter 2.5.2. An energy-loss window of 50 eV with an edge onset at 880 eV was used for the Ce-$M_{4,5}$ ionization edge. For the Sm-$M_{4,5}$ edge with the onset at 1070 eV an energy-loss window of 60 eV was used. The convergence angle of the electron probe was 20.1 mrad. The collection angle was set to 15.5 mrad. The spectrometer dispersion of 0.5 eV/channel (with 2048 channels) was chosen to allow the simultaneous acquisition of the elemental ionization edges of Ce and Gd/Sm.

The EELS reciprocal sensitivity factor $\dfrac{1}{s_{Sm,Ce}}$ was determined in three different sample regions (see Fig. 4.1 in Chapter 4.2: RV, RL1, and RL2) with constant Sm- and Ce-concentrations. The concentrations of these regions were derived on the basis of quantitative EDXS analyses described in Chapter 2.4. EELS linescans across the RLs 1 and RLs 2 of Sm-1175 and Sm-1219 and the RVs of Sm-1175-20h were used to determine $s_{Sm,Ce}$. The mean concentration ratio $\dfrac{C_{Ce}}{C_{Sm}}$ determined by EDXS was calculated for the linescans across the RLs for the region with constant composition and then divided by the mean intensity ratio $\dfrac{J_{Ce}}{J_{Sm}}$ to calculate the value of $\dfrac{1}{s_{Sm,Ce}}$. In the case of the RVs only the data points with a Sm-concentration of about 80 at% were used to derive the sensitivity factor. The mean Sm-concentration of the linescans and the mean net intensities for the different samples and sample regions are shown in Tab. 2.3.

The values of $\frac{1}{s_{Sm,Ce}}$ from the different samples and sample regions show good agreement. The maximum deviation of the mean reciprocal sensitivity factor 1.68 in percent is 3.1 %. This value was used to quantify all electron energy-loss spectra. This procedure led in general to consistent compositions for both EDXS and EELS analyses.

3 The $Gd_xCe_{1-x}O_{2-x/2}$ System: Phase Formation and Cation Interdiffusion

The following chapter contains the results on the Gd_2O_3 thin films on polycrystalline ceria substrates. The samples were prepared to investigate microstructure evolution and cation interdiffusion at the Gd_2O_3/CeO_2 interface. Structure identification was performed and compared to structural findings of previous studies. Subsequently, concentration profiles across Gd_2O_3/CeO_2 interfaces were measured to investigate the miscibility of Gd in CeO_2 and to determine the cation interdiffusion coefficient at temperatures ranging from 986 °C to 1175 °C. This allowed the calculation of the activation enthalpy of interdiffusion using an Arrhenius type relation.

3.1 Specimen Fabrication

3.1.1 Substrate Preparation

Polycrystalline CeO_2 substrates were prepared by employing a multi-step procedure. High-purity CeO_2 powder was purchased from Treibacher Industrie AG (Althofen, Austria). The amount of CeO_2 divided by the Total Rare Earth Oxides (TREO) is above 99.99 %. Given contents of Y_2O_3, LaO_3, Pr_6O_{11} and Nd_2O_3 are less than 10 ppm each. Contamination concentrations in the powder are 20 ppm for Si and below 20 ppm for Al, Fe, Sn, and Zn. The particle size distribution provided by Treibacher was that the diameter of 10 % of the particles is smaller than 0.136 µm and the diameter of 10 % of the particles is larger than 2.26 µm where the percentage refers to volume. Consolidation of the CeO_2 powder into cylindrical shape was performed by uniaxial pressing (10 kN) using a steel powder compaction tool with a diameter of 20 mm (resulting pressure 31, 8 MPa). An amount of 4.5 g for each pellet resulted in green bodies, about 4 mm in thickness. The relative density (50 %) was estimated by the mass and the geometrical parameters of the green body. To enhance the green-body den-

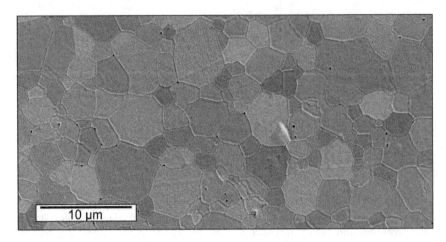

Figure 3.1: SEM image of densified CeO_2 used as substrate, the sample was thermally etched for better visibility of the grain boundaries.

sity prior to sintering, Cold Isostatic Pressing (CIP) was performed at 400 MPa resulting in a relative density of 65 %.

Pure CeO_2 was observed to densify poorly during sintering. To obtain bulk CeO_2 with a small degree of porosity, a Hot Isostatic Pressing (HIP) step was introduced in the fabrication process of the substrate pellets. The green bodies were pre-sintered 1350 °C for 2 h under static air using a Nabertherm HT 08/16S oven with heating and cooling rates of 5 $Kmin^{-1}$ to relative densities of 93 %, where only closed porosity remains in the discs. Subsequently HIP was performed in Ar-atmosphere using an ASEA QIH-6 hot isostatic press. To prevent contamination of the substrates in an Al_2O_3 crucible, the pre-sintered discs were embedded in CeO_2 powder. The substrates were heated to 1350 °C at a rate of 5 $Kmin^{-1}$. After reaching the maximum temperature an Ar pressure of 150 MPa was applied for 2 h. This induces the reduction of CeO_2 due to the low oxygen partial pressure in the inert Ar environment. Finally, the samples were heat treated at 900 °C in static air for 1 h for reoxidation. This multi-step fabrication of pure CeO_2 substrate pellets resulted in a final relative density of about 99.6 %. The cylindrical samples were cut using a diamond-wire saw into 3 substrates (about 1 mm thick). The substrates were polished (felt 1 µm) and then annealed at 900 °C for 24 h in static air to reduce the density of near-surface defects in the substrates. Fig. 3.1 shows a SEM image of a finished substrate after thermal etching. Grain sizes range from 1 µm to 8 µm. The dark spots correspond to pores.

Table 3.1: Denotation of the heat-treated Gd-samples

Sample	Temperature [°C]	Duration [h]
Gd-986	986	100
Gd-1069	1069	100
Gd-1122	1122	100
Gd-1175	1175	100
Gd-1270	1270	100

3.1.2 Thin Film Deposition

Gd_2O_3 layers were deposited on the polished CeO_2 substrates by Pulsed Laser Deposition (PLD). A Gd_2O_3 target was prepared from powder supplied by Alfa Aesar (Ward Hill, Massachusetts, USA) with Gd_2O_3/TREO > 99.99 % and the following impurity concentrations: 20 ppm for Y, 10 ppm for Ca, K, Si, Sm and less than 2 ppm for other elements. Targets were produced from the powder by uniaxial pressing and subsequent CIP as outlined before. This was followed by sintering at 1150 °C for 2 h and at 1450 °C for 2 h in a Nabertherm HT 08/16S oven with a heating and cooling rate of 5 Kmin^{-1}. PLD was carried out using a KrF-excimer laser with a wavelength of 248 nm from Lambda-Physik Göttingen (now Coherent, Santa Clara, California, USA). The substrate was heated to 500 °C with 5 Kmin^{-1} heating and 3 Kmin^{-1} cooling rate and positioned in a distance of about 5 cm from the target. The O_2 pressure in the deposition chamber was reduced to 6 Pa. 3500 pulses with a pulse energy of 200 mJ and a repetition rate of 2 Hz were used which resulted in the formation of an amorphous Gd_2O_3 layer with about 150 nm thickness.

3.1.3 Thermal Treatments

Annealing treatments were carried out for 100 h at temperatures of 986 °C (sample denotation Gd-986), 1069 °C (Gd-1069), 1122 °C (Gd-1122), 1175 °C (Gd-1175), and 1270 °C (Gd-1270) in a tube furnace (model HTRH 40-100/16 from GERO Hochtemperaturöfen GmbH & Co. KG, Neuhausen, Baden-Württemberg, Germany). The heating and cooling rates were 5 Kmin^{-1} for all samples. At the end of the cooling process the cooling rate was even lower. Annealing was performed in an oxidizing atmosphere with a gas flux of 0.8 lmin^{-1} Ar and 0.2 lmin^{-1} O_2. The temperature range was chosen to yield interdiffusion profiles broad enough for investigation. Sample denotations with the corresponding annealing temperature and duration are listed in Tab. 3.1.

3.1.4 Sample Preparation for Transmission Electron Microscopy

Cross-section TEM samples were prepared from all specimens by cutting thin slices with a precision diamond wire saw (well Diamantdrahtsägen GmbH, Mannheim, Baden-Württemberg, Germany) which were glued together (M-Bond 610, Vishay Micro-Measurements, Vishay Precision Group, Wendell, North Carolina, USA) with the Gd_2O_3-dopant film facing against each other. The resulting cuboid (about 2 mm thick) was grinded to a resulting total thickness of 390 μm from both sides with a PHOENIX 4000 grinding machine system from Buehler (Düsseldorf, Nordrhein-Westfalen, Germany) equipped with diamond abrasive paper (30 μm grain size). The flat sandwich was then fixed in a slit in an Al_2O_3 tube with 3 mm diameter using G2 Epoxy glue from Gatan (curing 160 °C). Thin slices were cut from the tube utilizing a wire saw type WS-22 from K. D. Unipress (Princeton, New Jersey, USA) with a wire thickness of about 60 μm and diamond wire-saw suspension 6 μm from Oberflächentechnologien Ziesmer, Germany, Bavaria, Germany). The resulting discs have thickness of about 300 μm. After double side face grinding to a thickness of 80 μm, additional grinding and polishing (3/1/0.5 μm-METADI suspension, Buehler) of the central part of the disc down to $4 - 6$ μm was carried out with a Gatan Model 656 Dimple grinder. A Gatan Model 691 Precision ion polishing system was used for single-sector Ar^+-ion milling with an ion energy of 3.0 kV as described by Dieterle et al. [166]. The finished electron transparent sample was carbon-coated to prevent charging in the transmission electron microscope. To reduce contamination in the transmission electron microscope, the samples were cleaned in a plasma cleaner model TPS 216 EC from Binder Labortechnik, Hebertshausen, Bavaria, Germany (30 s, Ar-plasma).

3.2 Microstructure Characterization

The microstructural changes at the Gd_2O_3/CeO_2 interface were studied in dependence of the annealing temperature to study the microstructural evolution. The interface between Gd_2O_3 and CeO_2 is of particular interest with respect to cation interdiffusion studies (Chapter 1.4.2). The annealing conditions must be chosen to ensure a direct interface between the materials.

Cross-section BFTEM images of the Gd_2O_3/CeO_2 heterostructure are shown in Fig. 3.2 to illustrate the microstructural evolution of the samples in dependence of the annealing temperature. The amorphous Gd_2O_3 thin film on the CeO_2 substrate present before the anneal is shown in Fig. 3.2a) which reveals a homogeneous layer thickness of 150 nm with a flat interface between

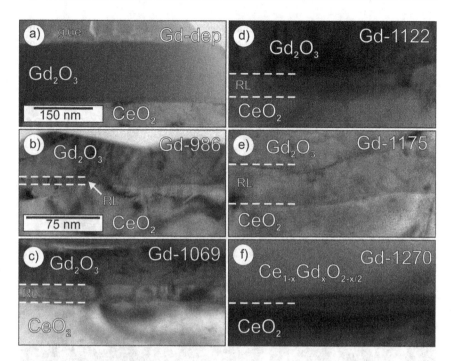

Figure 3.2: Cross-section BFTEM images of the interface region between Gd_2O_3 and CeO_2 a) as-deposited sample before annealing and samples after annealing for 100 h at b) 986 °C, c) 1069 °C, d) 1122 °C, e) 1175 °C, and f) 1270 °C. The size marker in b) also applies to c-f).

the Gd_2O_3 thin film and the CeO_2 substrate. For the annealed samples in the temperature range from 986 °C to 1175 °C (Fig. 3.2b-e)) two well-defined layers can be recognized on the CeO_2 substrate. The first one, indicated by the dashed lines, can be observed between the substrate and the remaining Gd_2O_3 thin film (the second one). This RL increases in thickness with increasing temperature from about 8 nm in Gd-986 over 15 nm in Gd-1069 and 24 nm in Gd-1122 up to about 40 nm in Gd-1175. Increasing the annealing temperature to 1270 °C leads to the complete consumption of the Gd_2O_3 film (Fig. 3.2f)).

The crystalline structure in the different regions (CeO_2 substrate, Gd_2O_3 layer, $Gd_xCe_{1-x}O_{2-x/2}$ RL) was analyzed by HRTEM. We note that SAED patterns could not be taken due to the small size of the regions to be analyzed. The analysis procedure is exemplarily illustrated for sample Gd-1122. Fig. 3.3 shows a HRTEM cross-section image of the interface region comprising the

CeO_2 substrate, RL, and Gd_2O_3. Small sections of the different sample regions, indicated by white frames, were selected for structure analyses. The Fourier-transforms (diffractograms) of these regions are shown in Figs. 3.4a), c), and e) with matching simulated diffraction patterns in Figs. 3.4b), d), and f).

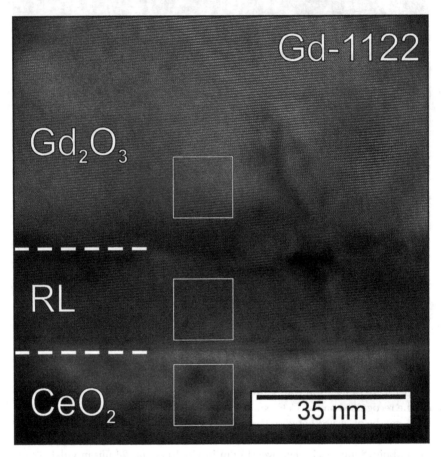

Figure 3.3: HRTEM cross-section image of Gd-1122 with white frames indicating regions of CeO_2, Gd_2O_3 and the RL. The Fourier-transforms of these regions and the corresponding simulated diffraction patterns are shown in Fig. 3.4.

The diffractogram in Fig. 3.4a) was obtained from the CeO_2 substrate. Simulations of diffraction patterns were performed with the software package JEMS version 3.3826U2009 by Stadelmann [136]. The corresponding simula-

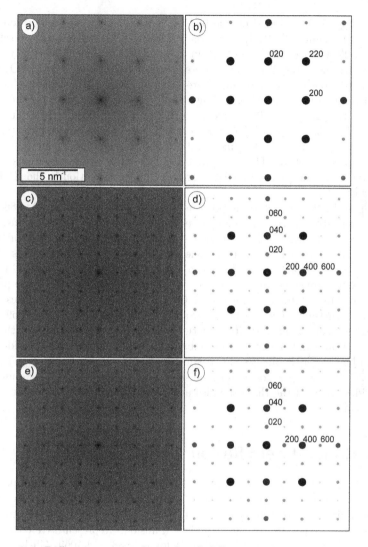

Figure 3.4: Diffractograms and simulated diffraction patterns of different regions of Gd-1122 (see insets in Fig. 3.3) with a) diffractogram of CeO_2 substrate, b) simulated diffraction pattern for CeO_2 with fluorite structure along the $[0\,0\,1]$ zone axis, c) diffractogram of the RL, d) simulated diffraction pattern of $Gd_{0.6}Ce_{0.4}O_{1.7}$ with bixbyite structure along the $[0\,0\,1]$ zone axis, e) diffractogram of Gd_2O_3 film, and f) simulated diffraction pattern of Gd_2O_3 with bixbyite structure along the $[0\,0\,1]$ zone axis.

tion in Fig. 3.4b) is based on the fluorite structure (crystal structure data in Tab. 1.1) assuming a [0 0 1] zone axis orientation. As expected, the positions of the reflections in the diffractogram and simulation coincide perfectly. The diffractograms in Fig. 3.4c) and e) were obtained from the RL and Gd_2O_3 film. The simulations in Fig. 3.4d) and f) are based on the bixbyite structure (crystal structure data in Tab. 1.1) assuming a [0 0 1] zone axis orientation. The occupancy of the cations is 0.6 for the Gd- and 0.4 for the Ce-ions for the simulations of the RL ($Gd_{0.6}Ce_{0.4}O_{1.7}$). The positions of the reflections of the diffractogram and simulations agree well demonstrating that RL and Gd_2O_3 are present in the bixbyite structure. The high intensity of the $\langle n\,0\,0\rangle$-type reflections in the diffractogram can be attributed to nonlinear image formation in the HRTEM image Fig. 3.3.

Crystal structure determination was analogously performed for Gd-986, Gd-1069, and Gd-1175. HRTEM images were acquired and their Fourier transforms of the different areas of interest were compared to simulations. All following simulations correspond to a [1 0 1] zone axis orientation. Diffractograms of the substrate region of Gd-986 (Fig. 3.5a)), Gd-1069 (Fig. 3.5b)), and Gd-1175 (Fig. 3.54c)) are shown in the first row of Fig. 3.5. As expected, by comparing to a simulated diffraction pattern of CeO_2 (Fig. 3.5d)), the substrates have fluorite structure. The Gd_2O_3 films and the RLs of Gd-986, Gd-1069, and Gd-1175 occur in the bixbyite structure. The reflection positions of the RL diffractograms and the simulated bixbyite structure (Fig. 3.5h)) coincide well as can be seen in the images of Gd-986 (Fig. 3.5e)), Gd-1069 (Fig. 3.54f)), and Gd-1175 (Fig. 3.5g)). Concluding the structural analysis the same is found for the Gd_2O_3 film of the three samples again comparing the reflection positions (Fig. 3.5i-l)).

3.3 Cation Interdiffusion

EDXS line profiles were performed to determine concentration profiles perpendicular to the Gd_2O_3/CeO_2 interface. Composition profiles were not measured in the vicinity of grain boundaries in the Gd_2O_3 layer and CeO_2 substrate to avoid contributions from grain boundary diffusion. Representative concentration profiles are shown in Fig. 3.6 a–d) for Gd-986, Gd-1069, Gd-1122, and Gd-1175. A concentration profile for Gd-1270 is not presented because the concentration of the remaining $Gd_xCe_{1-x}O_{2-x/2}$ thin film is laterally inhomogeneous. The concentration profiles could be well fitted with Eq. 1.8. This equation corresponds to the solution of Fick's second law with concentration-independent interdiffusion coefficient D for a diffusion couple, assuming an initial Gd-concentration $C_0 = 100$ at% at $t = 0$ for a position $x < 0$ and

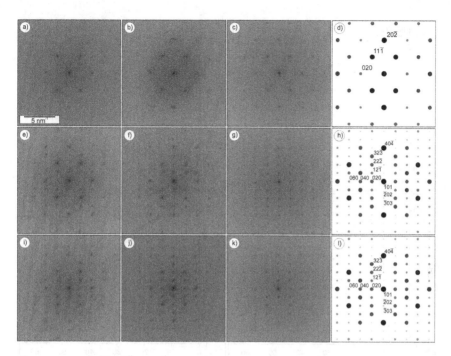

Figure 3.5: Diffractograms and simulated diffraction patterns of different regions of Gd-986, Gd-1069, and Gd-1175. Diffractograms of the CeO_2 substrate are presented for a) Gd-986, b) Gd-1069, c) Gd-1175, of the RL for e) Gd-986, f) Gd-1069, g) Gd-1175, of the Gd_2O_3 thin film for i) Gd-986, j) Gd-1069, k) Gd-1175, and simulated diffraction patterns for d) CeO_2 with fluorite structure, h) $Gd_{0.6}Ce_{0.4}O_{1.7}$ with bixbyite structure l) Gd_2O_3 with bixbyite structure all simulated along the $[1\,0\,1]$ zone axis.

$C_0 = 0$ at% for $x > 0$ at $t = 0$ with an interface position at $x = 0$ as discussed in Chapter 1.3.1. For the experimental profiles x was substituted by $x' - d$ with a fit parameter d to move the function along the position-axis. Based on the fit curves for the experimental diffusion profiles, diffusion coefficients were determined. A minimum of 30 concentration profiles were evaluated for each temperature resulting in the following averaged interdiffusion coefficients: $6.53 \cdot 10^{-20} \pm 5.62 \cdot 10^{-21}$ cm²s⁻¹ for Gd-986, $1.63 \cdot 10^{-19} \pm 1.32 \cdot 10^{-20}$ cm²s⁻¹ for Gd-1069, $4.7 \cdot 10^{-19} \pm 4.6 \cdot 10^{-20}$ cm²s⁻¹ for Gd-1122, and $1.48 \cdot 10^{-19} \pm 0.12 \cdot 10^{-19}$ cm²s⁻¹ for Gd-1175. The error margins comprise the

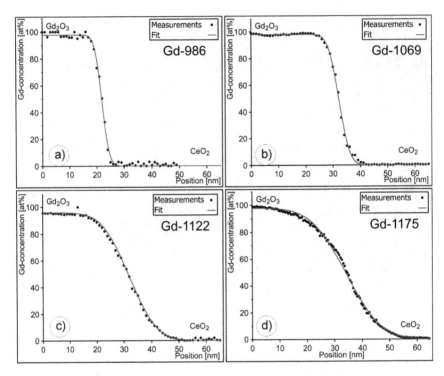

Figure 3.6: Gd-concentration profiles perpendicular to the Gd_2O_3/CeO_2 interface obtained by quantitative EDXS measurements (black dots) with fit curves based on Eq. 1.8 (red lines) for a) Gd-986, b) Gd-1069, c) Gd-1122, and d) Gd-1175.

statistical error and the error due to possible deviations from the intended annealing temperature.

3.4 Discussion

The presented results are discussed at first with regard to the phase diagram of the $Gd_xCe_{1-x}O_{2-x/2}$ system and then, secondly, the interdiffusion of the cations in dependence of temperature is assessed.

3.4.1 Phase Formation in the $Gd_xCe_{1-x}O_{2-x/2}$ System

Crystal structure analyses confirm the fluorite structure for CeO_2 for all investigated samples in the temperature range between 986 °C and 1175 °C as anticipated from literature [24]. Formation of discrete stoichiometries with different crystal structures was observed at low O partial pressures in the temperature range between 600 °C and 1500 °C [68–70] (see Chapter 1.2.1). This was checked in this work to ensure that even at low temperatures the low oxygen partial pressure in the electron microscope does not lead to reduction of the CeO_2. However, dark spots in the substrate regions of the Gd-samples in Fig. 3.2 suggest that reduction under electron radiation occured and produced defects as reported by Yasunaga et al. [71]. Crystal structure analyses of the remaining pure Gd_2O_3 film as well as the RLs yield the bixbyite structure in the temperature range between 986 °C and 1175 °C. At 1270 °C, the Gd_2O_3 film is completely consumed leaving a $Gd_xCe_{1-x}O_{2-x/2}$ layer with laterally and vertically varying composition.

Different crystal structures were reported for pure Gd_2O_3. Numerous studies tried to determine the transition temperature between the bixbyite (low-temperature phase) and monoclinic structure (high-temperature phase) experimentally [65, 87–94]. The change from the bixbyite structure to the monoclinic structure was determined at temperatures ranging from 800 °C to 1400 °C (see Fig. 1.1). More recently Zinkevich summarized known data [56] and suggested a transition temperature of 1152 ± 20 °C which seems to be an acceptable estimation. The meaurements which show a monoclinic phase below 1100 °C are probably due to contamination of the Gd_2O_3 with different REs for the earlier studies. Additionally, if a monoclinic→bixbyite transition was found at temperatures below 1100 °C, the experimental conditions provided an alternative transformation path involving water, high pressure, or special starting material which may indicate that the bixbyite structure is metastable and the monoclinic structure is the equilibrium structure not only at high, but also at low temperatures.

In our study, the Gd_2O_3 film was found to be present in the bixbyite structure up to 1175 °C. This is consistent with references [56, 93] considering the given error margin. Some of the discrepancies in the literature data can be attributed to the fact that thermodynamic equilibrium may not have been achieved in some cases as discussed in Chapter 1.2.2. Typically, bulk samples are sintered at high temperatures during the fabrication process and long annealing times are necessary to establish thermodynamic equilibrium at lower temperatures. With respect to phase formation it is noted that Gd_2O_3 in our samples is originally present as a thin film with amorphous structure after PLD. Hence, equilibrium phase formation will be less affected by the slow kinetics of transitions between different crystalline phases because the thermodynamically stable phase is most likely formed from the amorphous state. This indicates that the bixbyite structure is not metastable and the monoclinic phase does not occur at thermal equilibrium at low temperatures.

Composition profiles across the Gd_2O_3/CeO_2 interface of Gd-986, Gd-1069, Gd-1122 and Gd-1175 shown in Fig. 3.6a-d) in Section 3.3 continuously cover the complete concentration range of $Gd_xCe_{1-x}O_{2-x/2}$. Indications for a miscibility gap, such as a step in the concentration profile as shown in Chapter 1.3.2, were not observed in the whole studied temperature range between 986 °C and 1175 °C. Comparing the width of concentration profiles in Fig. 3.6 and the RLs in Fig. 3.2 shows that the width of the RLs is smaller than the width of the concentration profiles. This suggests that $Gd_xCe_{1-x}O_{2-x/2}$ with small Gd concentrations is present in the fluorite structure whereas the bixbyite structure prevails in the RLs. The exact composition for the structure transition between fluorite and bixbyite could not be determined because the EDXS line profiles cannot be superimposed on the BFTEM images with sufficient precision. Information on the phase diagram of the $Gd_xCe_{1-x}O_{2-x/2}$ system is only available for temperatures above 1200 °C.

Previous studies disagree regarding the existence of a miscibility gap between the fluorite and the bixbyite phases in the $Gd_xCe_{1-x}O_{2-x/2}$ system. Brauer and Gradinger [72] and Grover and Tyagi [74] did not find a miscibility gap at 1400 °C. In contrast, Bevan et al. [57, 104] proposed a miscibility gap between the fluorite and bixbyite phases for Gd concentrations between 54 at% and 74 at% at 1600 °C. Using synchrotron powder X-ray diffraction at 1200 °C, Artini et al. [99] did not observe a two-phase region over the whole composition range. They propose a gradual transition between the fluorite and the bixbyite phase by the formation of microdomains of Gd_2O_3, which are formed in the fluorite matrix. Zinkevich [97] proposed a complete phase diagram based on calculations, measurements and experimental data. According to this work, a large miscibility gap between the fluorite and bixbyite phases is predicted between 900 °C to 1200 °C. According to Ye et al. [100], Li et al. [101], and Ye et al. [102] a gradual transition from the fluorite into the bixbyite structure occurs in the $Gd_xCe_{1-x}O_{2-x/2}$ system starting at a Gd concentration of 10 at%. The transition is facilitated by the formation of defect clusters, domains and bixbyite precipitates with increasing Gd concentration due to the ordering of aggregated cations and oxygen vacancies. The results are consistent with Ref. [99] and Refs. [100–102] which propose a gradual transition between the fluorite and bixbyite phases. The composition profiles across the Gd_2O_3/CeO_2 interface (Fig. 3.6) indicate full miscibility in the temperature interval between 986 °C and 1175 °C. We did not observe nano-sized domains which are, however, difficult to observe in the thin RLs of our samples. Moreover, ordering of vacancies and, hence, cluster formation may be inhibited in an environment, where inter-diffusion takes place.

The absence of a miscibility gap at 986 °C and broad single phase region across the complete concentration range of $Gd_xCe_{1-x}O_{2-x/2}$ are properties which make further investigation and developement of the material system as solid electrolyte worthwhile. The absence of a miscibility gap in GDC at high tem-peratures is promising with respect to long-term stability of the material in ap-plications, e.g, as solid electrolyte in SOFCs. The formation of defect clusters at low Gd-concentrations of 10 at% may affect the oxygen conductivity of the ma-terial, but does not change the crystal structure at this composition [100–102]. In comparison to YDZ, the stability of the material is not limited by decom-position. This allows to focus research efforts on material compatibility issues with other materials.

3.4.2 Interdiffusion at the Gd_2O_3/CeO_2 Interface

The acquired EDXS line profiles were used to investigate the interdiffusion pro-cess at the Gd_2O_3/CeO_2 interface. In general, composition-dependent interdif-fusion coefficients must be assumed which manifest themselves by asymmetrical concentration profiles. The high symmetry of the profiles in Fig. 3.6 and the good fit of the experimental data by Eq. 1.8 indicate that composition de-pendence is not pronounced in the analyzed system and temperature range. For comparison, Boltzmann-Matano analyses were performed which yield the same interdiffusion coefficients in a wide concentration range disregarding Gd-concentrations below 20 at% and above 80 at% which are prone to large errors using this method.

An Arrhenius-type temperature dependence of the diffusion coefficient ac-cording to Eq. 1.13 (see Chapter 1.4.1) is assumed with the concentration-independent interdiffusion coefficient D, the frequqency factor D_0, the activa-tion enthalpy ΔH, the Boltzmann constant k_B, and the temperature T. A plot of the logarithm of the diffusion coefficients as a function of the reciprocal temperature is shown in Fig. 3.7. A straight line can be well fitted to the data with a slope which yields an activation enthalpy of 2.29 ± 0.22 eV/atom and a frequency factor of $9.09 \cdot 10^{-11} \pm 4.18 \cdot 10^{-12}$ cm^{2}s^{-1}. The error is due to the statistical error for D values, the linear regression, and the temperature uncer-tainty. This activation enthalpy is characteristic to bulk interdiffusion across the Gd_2O_3/CeO_2 interface. The interdiffusion coefficient determined in this work is the first value from diffusion couple experiments.

Previous data is only available from grain growth data, which yield a wide range of values. The lower activation enthalpy of 1.32 eV/atom for grain coars-

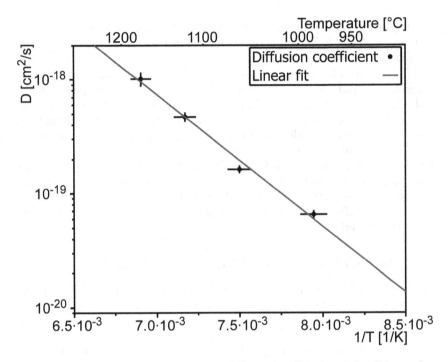

Figure 3.7: Arrhenius plot of the interdiffusion coefficients of the Gd-samples with a linear fit curve to derive the activation enthalpy for interdiffusion

ening in dense nanocrystalline $Gd_{0.22}Ce_{0.78}O_{1.89}$ thin films [130] was attributed to the fact that grain boundary diffusion mainly contributes to the coarsening process. High activation enthalpies between 4.5 eV/atom and 9 eV/atom were determined for bulk-sintered Gd-doped ceria [126, 127, 129]. Different models were used to assign these values to either to bulk interdiffusion or grain boundary diffusion. However, it was not possible to differentiate the diffusion processes in all studies coherently. The acquired activation enthalpies are high compared to the 2.29 eV/atom obtained in our study indicating that cation transport across grain boundaries is impeded during grain growth, e.g., by segregation and the presence of a grain boundary phase [126] and that other diffusion processes like surface diffusion may play a role during the grain growth process. The latter effects can be excluded in our study. Due to experimental limitations, diffusion coefficients could be only determined for four different temperatures. However, the experimental data can be well fitted by a straight line with a relatively small error for ΔH and D_0, which supports the validity of the results despite the small number of data points. It is noted that the extrapolation of the interdiffusion coefficients toward higher temperatures on the basis of the determined values of ΔH and D_0 needs be carried out with care if the structure of the Gd_2O_3 changes from the bixbyite into the monoclinic crystal structure.

4 The $Sm_xCe_{1-x}O_{2-x/2}$ System: Phase Formation and Cation Interdiffusion

This chapter contains the results for the Sm_2O_3 thin films on CeO_2 substrates. Microstructural evolution and cation interdiffusion of the SDC system were investigated. Crystal structure identification in the different samples and concentration profiles reveal a more complex phase evolution than previously found for the GDC samples. The acquired concentration profiles nevertheless allow to determine cation interdiffusion coefficients in the temperature range from 987 °C to 1175 °C. The activation enthalpy for interdiffusion can be calculated assuming an Arrhenius-type relation.

4.1 Specimen Fabrication

The same fabrication procedures were used for sample preparation as for the preparation of the Gd samples described in detail in Chapter 3.1. Deviations from this process are described in the following.

4.1.1 Substrate Preparation

Pressing and sintering of the CeO_2 substrate was performed in exactly the same way as for the Gd samples (see Chapter 3.1.1).

4.1.2 Thin Film Deposition

Instead of Gd_2O_3 layers, Sm_2O_3 layers were deposited on the polished CeO_2 substrates by PLD. A Sm_2O_3 powder from Alfa Aesar (Sm_2O_3/TREO > 99.99 %, impurity concentrations: 20 ppm for Y, 10 ppm for Ca, Gd, K, Si and less than 2 ppm for other elements) was used. The process of producing the targets and the PLD parameters were the same as presented in Chapter 3.1.2. The resulting film thickness was approximately 180 nm.

Table 4.1: Denotation of the heat-treated Sm samples

Sample	Temperature [°C]	Duration [h]
Sm-987	987	100
Sm-1073	1073	100
Sm-1175	1175	100
Sm-1219	1219	100
Sm-1266	1266	100
Sm-1122-40h	1122	40
Sm-1175-20h	1175	20
Sm-1175-50h	1175	50

4.1.3 Thermal Treatments

Annealing treatments were carried out with the same heating and cooling rates and gas fluxes as given in Chapter 3.1.3. Samples were heat treated for 100 h at temperatures of 987 °C (sample denotation Sm-987), 1073 °C (Sm-1073), 1175 °C (Sm-1175), 1219 °C (Sm-1219), and 1266 °C (Sm-1266). Additional heat treatments with shorter annealing times were performed to investigate the temporal evolution of the sample interfaces for 40 h at 1122 °C (Sm-1122-40h), for 20 h at 1175 °C (Sm-1175-20h), and 50 h at 1175 °C (Sm-1175-50h). The temperature range was chosen to yield interdiffusion profiles broad enough for investigation. The denotation of the samples and the corresponding annealing temperature and duration are listed in Tab. 4.1.

4.1.4 Sample Preparation for Transmission Electron Microscopy

No changes were applied to the preparation routine used for the Gd samples (see Chapter 3.1.4).

4.2 Microstructural Characterization

The cross-section BFTEM images in Fig. 4.1 illustrate the structural evolution of the specimens as a function of annealing temperature and annealing time. Fig. 4.1a) shows the amorphous as-deposited Sm_2O_3 film with a homogeneous thickness of approximately 180 nm on the CeO_2 substrate. Typical grain sizes in the CeO_2 substrate are between 1 μm and 8 μm. Upon annealing at 987 °C for 100 h (Fig. 4.1b)), the Sm_2O_3 layer completely crystallizes and is clearly distinguishable from the CeO_2 substrate. Typical grain sizes at this tempera-

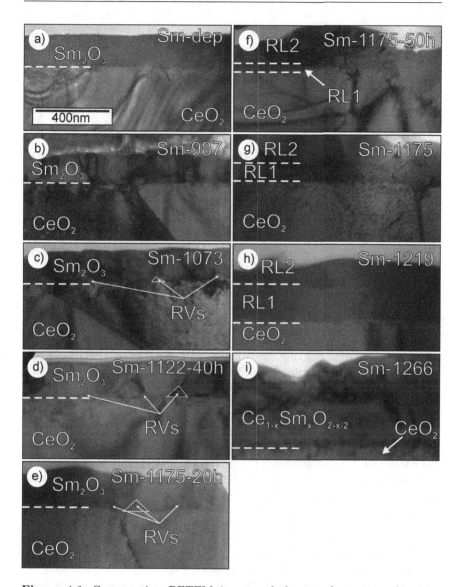

Figure 4.1: Cross-section BFTEM images of the interface region between Sm_2O_3 and CeO_2: a) as-deposited sample before annealing and samples after annealing at b) 986 °C for 100 h, c) 1069 °C for 100 h, d) 1122 °C for 40 h, e) 1175 °C for 20 h, f) 1175 °C for 50 h, g) 1175 °C for 100 h, h) 1219 °C for 100 h, and i) 1266 °C for 100 h.

ture are between 80 nm and 180 nm. Larger grains at higher temperatures are indicative of significant grain boundary motion.

First structural changes in the Sm_2O_3 layer become recognizable at higher temperatures as shown for Sm-1073, Sm-1122-40h, and Sm-1175-20h (Figs. 4.1c)-e)). In all three samples triangular-shaped regions with slightly different contrast are visible at the intersections of grain boundaries of the Sm_2O_3 film and the Sm_2O_3/CeO_2 interface. These regions are referred to as Reaction Volume (RV).

A continuous Reaction Layer 1 (RL1) is formed upon longer annealing at 1175 °C in Sm-1175-50h marked by dashed lines in Fig. 4.1f) with a second reaction layer on top (Reaction Layer 2 (RL2)). RL1 and RL2 differ in composition and crystal structure as demonstrated in the following. The thickness of RL1 increases from approximately 40 nm in Sm-1175-50h (Fig. 4.1f)) to about 100 nm in Sm-1175 (Fig. 4.1g)) and about 200 nm in Sm-1219 (Fig. 4.1h)). Annealing at 1266 °C (Fig. 4.1i)) leads to the formation of a single layer with inhomogeneous Sm- and Ce-concentrations.

The crystalline structure in the different sample regions (CeO_2 substrate, RVs, RL1, RL2, Sm_2O_3 film) was analyzed by SAED and HRTEM. SAED patterns and Fourier-transformed HRTEM images are compared with simulated diffraction patterns of the possible phases introduced in Chapter 1.2. The corresponding crystal structure data used for simulation is compiled in Tab. 1.1.

A representative example for structure identification by HRTEM is presented in Fig. 4.2 for sample Sm-1073. Fig. 4.2a) shows a HRTEM image of a sample region containing the CeO_2 substrate, two grains of the Sm_2O_3 film and a triangular-shaped RV at a Sm_2O_3 grain boundary. The CeO_2 substrate is oriented along the $[0\,0\,1]_f$ zone axis with the index f denoting the fluorite structure of CeO_2. Correspondingly, b1 and b2 refer to the bixbyite structure with the $Ia\bar{3}$ and $I2_13$ space groups, while m denotes the monoclinic structure. The Sm_2O_3 layer above the RV contains a small-angle grain boundary. Diffractograms of the four different regions indicated by black frames in the HRTEM image Fig. 4.2a) are shown in Figs. 4.2b-e). Corresponding simulated diffraction patterns are presented in Figs. 4.2f-i). As expected, the diffractogram of the CeO_2 substrate agrees with the simulated pattern of the fluorite structure (Figs. 2b,f). The diffractogram of the RV is explained by the bixbyite structure along $[0\,0\,1]_{b1}$ zone axis (Figs. 2c,g). It has to be noted here that the positions of the $\{2\,0\,0\}_f$ reflections of the fluorite structure coincide with the $\{4\,0\,0\}_{b1}$ reflections of the bixbyite structure due to the doubling of the lattice parameter in the bixbyite structure compared to the fluorite structure. Figs. 4.2d,e) show diffractograms of the two adjacent Sm_2O_3 grains oriented close to the $[1\,3\,0]_m$ zone axis together with the calculated diffraction patterns in Figs. 4.2h,i) which confirm

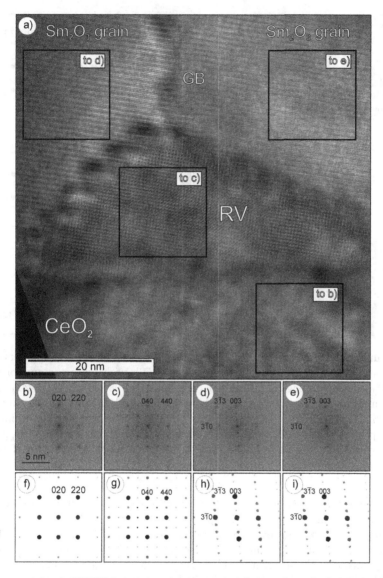

Figure 4.2: a) HRTEM cross-section image of Sm-1073 with the CeO_2 substrate oriented along the $[0\,0\,1]_f$. Diffractograms of different regions, indicated by black frames in the HRTEM image, with b) the CeO_2 substrate c) the RV along the $[0\,0\,1]_{b1}$, d) and e) two adjacent Sm_2O_3 grains along the $[1\,3\,0]_m$ zone axis. f-i) show simulated diffraction patterns of the respective crystal structures.

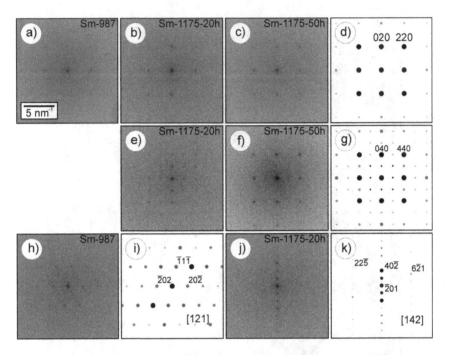

Figure 4.3: Diffractograms and simulated diffraction patterns of different regions of Sm-987, Sm-1175-20h, and Sm-1175-50h with diffractograms of the CeO_2 substrate along the $[0\,0\,1]_f$ zone axis of a) Sm-987, b) Sm-1175-20h, and c) Sm-1175-50h, of the RL1 of e) Sm-1175-20h, and f) Sm-1175-50h, and of the Sm_2O_3 thin film of h) Sm-987 and j) Sm-1175-20h, and simulated diffraction patterns of d) CeO_2 (fluorite structure, $[0\,0\,1]_f$ zone axis), g) $Sm_{0.6}Ce_{0.4}O_{1.7}$ (bixbyite structure, $[0\,0\,1]_{b1}$ zone axis), i) Sm_2O_3 (monoclinic structure, $[1\,2\,1]_m$ zone axis), and k) Sm_2O_3 (monoclinic structure, $[1\,4\,2]_m$ zone axis).

the monoclinic Sm_2O_3 structure. Missing reflections in the diffractogram compared to the simulated diffraction pattern can be explained by the tilt of the incident electron-beam direction from the perfect zone axis orientation. The observation of the monoclinic Sm_2O_3 structure, which is metastable at room temperature, is not surprising because the transition in the thermodynamically stable bixbyite structure is inhibited for kinetic reasons. The lattice planes of the RV continue across the different grain boundaries, which indicate partial structural coherence between the RV, Sm_2O_3 and CeO_2.

Crystal structure determination of Sm-987, Sm-1175-20, and Sm-1175-50h (with the exception of the RL2 on top of Sm-1175-50h) was performed as described for Sm-1073. The diffractograms are shown with corresponding simulations of SAED diffraction patterns in Fig. 4.3. Diffractograms of the substrate region of Sm-987 (Fig. 4.3a)), Sm-1175-20h (Fig. 4.3b)), and Sm-1175-50h (Fig. 4.34c)) are shown in the first row of Fig. 4.3. As expected by comparing to a simulated diffraction pattern of CeO_2 (Fig. 4.3d)), the substrates occur in the fluorite structure. The RL1s of Sm-1175-20h and Sm-1175-50h have bixbyite structure with the $Ia\bar{3}$ space group. The reflection positions of the RV and RL1 diffractograms and the simulated bixbyite structure (Fig. 4.3g)) coincide well as can be seen in the diffractograms of the RVs of Sm-1175-20h (Fig. 4.3e)) and the RL1 of Sm-1775-50h (Fig. 4.3f)). Sm-987 does not contain reaction phases. Hence, no further diffractograms are shown for this sample. The remaining Sm_2O_3 thin film of Sm-987 and Sm-1175-20h has monoclinic structure. A diffractogram of the Sm_2O_3 thin film of Sm-987 presented in Fig. 4.3h) is compared to a simulated diffraction pattern of monoclinic Sm_2O_3 along the $[1\,2\,1]_m$ zone axis in Fig. 4.3i). The reflection positions match well and confirm the monoclinic structure of the Sm_2O_3 film for Sm-987. The monoclinic structure was also found for the Sm_2O_3 film for Sm-1175-20h. A diffractogram of the Sm_2O_3 film of Sm-1175-20h is shown in Fig. 4.3j). The reflection positions in a simulated diffraction pattern of monoclinic Sm_2O_3 along the $[1\,4\,2]_m$ zone axis shown in Fig. 4.3k)) agree with experimental data.

Electron diffraction was used to determine the crystal structures of the continuous reaction layers RL1 and RL2 in Sm-1175-50h, Sm-1175, and Sm-1219. Representative SAED patterns taken from the three samples are displayed together with simulated ones in Fig. 4.4. The CeO_2 substrate was oriented in $[0\,0\,1]_f$ zone axis for SAED pattern acquisition. Experimental and simulated diffraction patterns of the CeO_2 substrate with fluorite structure are shown in Figs. 4.4a-c) for Sm-1175 and Sm-1219. For the RL1, good agreement between experimental (Sm-1175 and Sm-1219) and simulated SAED ($Sm_{0.6}Ce_{0.4}O_{1.7}$) diffraction patterns is found for the bixbyite structure with $Ia\bar{3}$ space group (Figs. 4.4d-f)). Structure identification of RL1 and substrate of Sm-1175-50h was carried out by HRTEM (see Fig. 4.3). The additional reflections in the patterns for the RL2 (Figs. 4.4g-i) of Sm-1175-50h, Sm-1175, and Sm-1219 are consistent with the $I2_13$ structure which was previously observed only for pure Sm_2O_3. Details (and differences) of the two structures are outlined in Chapter 1.2.3.

In addition to crystal structure determination, the investigation is complemented by chemical analyses. In Sm-987 only the fluorite and monoclinic phases of the pure materials were observed without obvious formation of any reaction

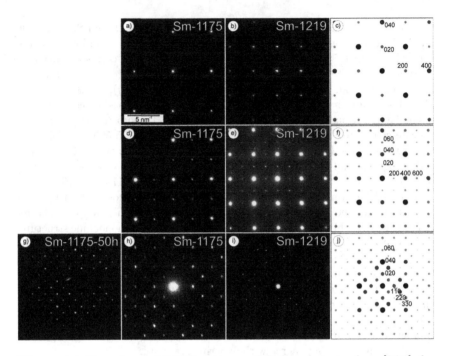

Figure 4.4: Experimental SAED patterns of CeO_2 substrate along $[0\,0\,1]_f$ for a) Sm-1175, b) Sm-1219, and c) simulated diffraction pattern along $[0\,0\,1]_f$ of the fluorite structure. Experimental SAED patterns of the RL1 along $[0\,0\,1]_{b1}$ for d) Sm-1175, e) Sm-1219, and f) simulated diffraction pattern along $[0\,0\,1]_{b1}$ of the diffraction patterns of the $Ia\bar{3}$ structure. Experimental SAED patterns of the RL2 along $[0\,0\,1]_{b2}$ for g) Sm-1175-50h, h) Sm-1175, and i) Sm-1219, and j) simulated diffraction pattern along $[0\,0\,1]_{b2}$ zone axis) of the $I2_13$ structure.

phase. This does of course not generally exclude reaction-phase formation at this temperature which was probably inhibited here due to the slow reaction kinetics (annealing time only 100 h). Formation of RVs with bixbyite structure ($Ia\bar{3}$ space group) is observed in Sm-1073, Sm 1122-40h, and Sm-1175-20h. Quantitative EDXS line-profile analyses across RV in these samples are presented in Figs. 4.5a-c) where the Sm-concentration is plotted as a function of the position x. The comparison of the graphs reveals that the central regions of the RVs exhibit a similar composition in spite of the specific thermal treatment, i.e., a local Sm-content of about 80 at% on the cationic sublattice.

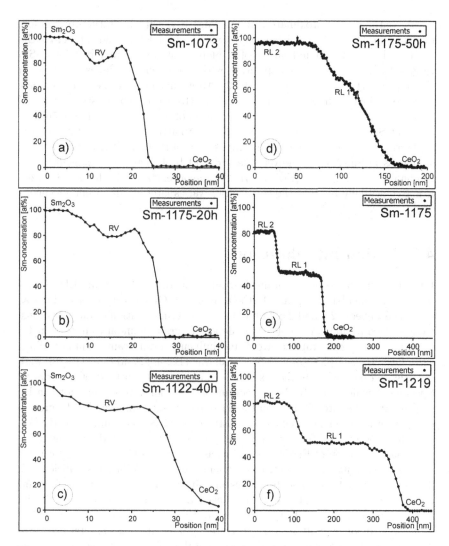

Figure 4.5: Sm-concentration profiles across RVs for a) Sm-1073, b) Sm-1175-20h, c) Sm-1122-40h, and Sm-concentration profiles across RL1 and RL2 for d) Sm-1175-50h, e) Sm-1175, and f) Sm-1219 obtained by quantitative EDXS and EELS measurements (black dots). Curves are guide to the eye.

A continuous, approximately 40 nm thick RL1 with $Ia\bar{3}$ structure was formed in Sm 1175-50h which correlates with a broad Sm-gradient across the reaction layer and an indication of a plateau at approximately 67 at% Sm (Fig. 4.5d)). A constant composition with about 95 at% Sm is observed in the upper part of the thin film in RL2. After annealing at the same temperature for a longer time (100 h in Sm-1175), RL1 with bixbyite structure ($Ia\bar{3}$) contains an almost constant Sm-concentration of about 50 at% (Fig. 4.5e)). A constant Sm-concentration of approximately 80 at% is found in RL2 with the $I2_13$ structure in this sample. The compositions of RL1 and RL2 in Sm-1219 are similar as in Sm-1175. However, the width of RL1 increases significantly due to ongoing interdiffusion (Fig. 4.5f)). A Sm-concentration profile for Sm-1266 is not shown here because the composition varies laterally in an inhomgeneous way.

4.3 Cation Interdiffusion

Interdiffusion profiles were obtained by quantification of EELS and EDXS linescans perpendicular to the Sm_2O_3/CeO_2 interface. Quantification procedures as explained in Secs. 2.4 and 2.5.2 were used to determine cationic concentrations. Composition profiles were not acquired close to grain boundaries in the Sm_2O_3 layer and CeO_2 substrate to avoid any influence from grain boundary diffusion. Only the samples Sm-987, Sm-1073, Sm-1122-40h, and Sm-1175-20h were investigated because only these samples contain direct Sm_2O_3/CeO_2 interfaces. Specimens, which already showed continuous reaction layers, were not considered because the interdiffusion in these samples is superimposed by the growth of the reaction layer into the CeO_2 substrate or the Sm_2O_3 film, respectively.

The experimental Sm-concentration profiles could be fitted with the diffusion couple solution of Fick's second law as outlined in Chapter 1.3.1. Eq. 1.8 was used assuming a concentration independent interdiffusion coefficient $\tilde{D}$. The experimental data and fit curves are shown in Fig. 4.6 for the four analyzed samples. The fit curves are based on the assumption of an initial Sm-concentration $C_0 = 100$ at% for a position $x < 0$ (initial Sm_2O_3 film) and $C_0 = 0$ for $x > 0$ (CeO_2 substrate) at $t = 0$ with an interface position at $x = 0$. To evaluate the experimental profiles, x was substituted by $x - d$ with a fit parameter d to shift the function along the position axis. Eq. 1.8 describes the shape of the experimental interdiffusion profiles very well. Thirty concentration profiles from different sample regions were evaluated for each sample by fitting Eq. 1.8 to the experimental data which yield average interdiffusion coefficients of $1.47 \cdot 10^{-20} \pm 2.31 \cdot 10^{-21}$ cm^2s^{-1} for Sm-987, $6.17 \cdot 10^{-20} \pm 9.24 \cdot 10^{-21}$ cm^2s^{-1} for Sm-1073, $1.58 \cdot 10^{-19} \pm 2.03 \cdot 10^{-20}$ cm^2s^{-1}

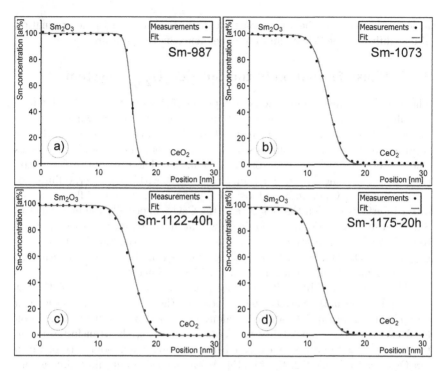

Figure 4.6: Sm-concentration profiles perpendicular to the Sm_2O_3/CeO_2 interface obtained by quantitative EDXS measurements (black dots) with fit curves based on Eq. 1.8 (red lines) for a) Sm-987, b) Sm-1073, c) Sm-1122-40h, and d) Sm-1175-20h.

for Sm-1122-40h, and $3.35 \cdot 10^{-19} \pm 3.10 \cdot 10^{-20}$ cm^{2}s^{-1} for Sm-1175. Only interdiffusion profiles with a maximal width of 6 nm were considered for Sm-987 because these profiles can be strongly affected by artificial broadening due to a tilt of the interface with respect to the incident electron beam. Hence, the interdiffusion coefficients for the temperature of 987 °C should be considered as an upper limit.

4.4 Discussion

The discussion of the Sm_2O_3/CeO_2 samples is subdivided in two parts. The first part covers the microstructural evolution and phase formation. Cation interdiffusion is discussed in the second part including the quantitative determination of interdiffusion coefficients and the dependence on temperature.

4.4.1 Phase Formation in the $Sm_xCe_{1-x}O_{2-x/2}$ System

The morphology and composition development of the investigated samples after annealing suggests that phase formation proceeds in the following way. To understand the difference in microstructure of the samples exhibiting RVs in contrast to those with continuous RLs, we point out that different processes lead to the final state of each of the specimens. Those comprise the initial crystallization of the Sm_2O_3 thin film, grain growth within the layer, diffusion of Ce along the Sm_2O_3 grain boundaries (and vice versa), and interdiffusion at the direct interface between the Sm_2O_3 and the CeO_2. Especially the movement of Sm_2O_3 grain boundaries due to grain growth at higher temperatures is expected to lead to the incorporation of Ce in the crystallites of the Sm_2O_3 film so that no pure Sm_2O_3 remains. This is observed for all samples, where a continuous RLs are formed during annealing (Sm-1175-50h – Sm-1219).

The formation of RVs with bixbyite phase is first observed in Sm-1073 at grain boundaries in the Sm_2O_3 films as shown in Figs. 4.1c-e). RV formation can be understood by rapid Ce-diffusion along grain boundaries in Sm_2O_3. Upon local excess of the Ce-solubility in monoclinic Sm_2O_3, the transformation in the bixbyite phase occurs as expected from the maximum solubility of Ce in Sm_2O_3 of $\approx$ 5 at% Ce [72, 97, 114, 115]. With increasing annealing time a continuous RL1 forms at the Sm_2O_3/CeO_2 interface in Sm-1175-50h (Fig. 4.1f)) which increases in width for Sm-1175 and Sm-1219 (Figs. 4.1g,h)). Continuous RL formation can be understood by the nucleation of the bixbyite phase at Sm_2O_3 grain boundaries and the simultaneous grain growth with its accompanied grain boundary movement in the Sm_2O_3 thin film. The Sm-concentration in the bixbyite RL1 decreases after formation from about 80 at% in the RVs to $\approx$ 50 at%, a composition that is compatible with the extrapolation of the phase diagram at higher temperatures (see phase diagram in Fig. 1.2) [72, 114]. Typically, a well-defined orientation relationship exists between RVs or RL1 and the CeO_2 substrate (Fig. 4.2) which can be understood by the similarity of the fluorite and bixbyite structure. In addition to the expected bixbyite phase (RL1) a second reaction layer (RL2) is detected on top of RL1 in Sm-1175-50h, where the monoclinic structure of Sm_2O_3 is transformed into the cubic $I2_13$ structure upon indiffusion of $\approx$ 5 at% Ce. Increasing the duration (Sm-1175) and temperature (Sm-1219) of the annealing treatment shows that RL2 remains in the $I2_13$ structure and the Sm-concentration stabilizes at $\approx$ 80 at%.

The detection of a second reaction layer (RL2) with the cubic $I2_13$ structure and Ce concentrations of 20 at% and 5 at% in addition to the cubic bixbyite phase with $Ia\overline{3}$ structure is unexpected because only the fluorite and $Ia\overline{3}$ crystal structures were observed up to now for $Sm_xCe_{1-x}O_{2-x/2}$ solid solutions [72, 97, 114, 115]. The $I2_13$ crystal structure was only reported for pure Sm_2O_3 between 620 °C and 730 °C [82] and for Sm_2O_3 films grown on cleaved NaCl crystal substrates at 250 °C and 300 °C [75, 106, 107]. The phase with $I2_13$ structure may be metastable at the high temperatures (1175 °C, 1219 °C) in our study. Its formation could be influenced by different factors. The presence of cubic bixbyite with $Ia\overline{3}$ structure and minimization of strain in thin-film samples could favor the appearance of the RL2 with $I2_13$ structure due to their structural similarity. Moreover, the presence of a strong oxygen-vacancy concentration gradient in the thin-film system and possible non-equilibrium O-concentrations on the anionic sublattice could be another reason for the appearance of the $I2_13$ structure. We note in addition that the similarity of the $Ia\overline{3}$ and $I2_13$ structures and lattice parameters requires a detailed structure analysis which could have prevented the detection of the phase with $I2_13$ structure in some previous studies. However, definite conclusions on the stability of the phase with $I2_13$ structure cannot be drawn without more detailed studies. Nevertheless, the absence of the monoclinic phase already at 5 at% Ce confirms the low Ce-solubility limit in the monoclinic structure.

Another unexpected observation concerns broad plateaus with constant Sm-concentration separated by steps in the composition profiles of Sm-1175 (Fig. 4.5e)) and Sm-1219 (Fig. 4.5f)). In interdiffusion profiles of diffusion couples, single-phase regions are generally characterized by a continuous composition change over their stability range as introduced in Chapter 1.3.2. Two-phase regions can be recognized by a vertical step across the composition range of the miscibility gap (depicted qualitatively in Fig. 1.3) [117]. The broad plateaus with almost constant composition between the steps in Figs. 4.5e,f) could be, on first sight, interpreted as single-phase regions with a narrow stability range. The steps between the plateaus suggest extended two-phase regions of the involved phases. An apparent broad miscibility gap between the fluorite and bixbyite phases is clearly unexpected with respect to previous results at higher temperatures. A miscibility gap between the fluorite and bixbyite phases was only reported at 1600 °C [97, 115], whereas single-phase regions were found at 1400 °C and 1280 °C [72, 114, 115]. A miscibility gap between the bixbyite and the monoclinic structure is likely due to the dissimilarity of the crystal structures and was indeed observed at 1280 °C, 1400 °C, and 1600 °C [97, 114, 115] at high Sm-concentrations > 80 at% except by Brauer and Gradinger [72].

To understand the observed composition profiles, the growth kinetics of reaction phases in thin-film diffusion couples has to be taken into account as outlined in Chapter 1.3.2. The growth kinetics of reaction phases is controlled by a combination of two processes, namely

- interdiffusion across the reaction phase.

- the rearrangement of the atoms at the interface which is necessary to form a new phase.

The latter process is denoted as reaction-controlled. For diffusion couples with sufficiently thick layers, interdiffusion governs the phase formation and typically all phases are found as predicted by the phase diagram. In contrast, diffusion is not necessarily rate-limiting anymore in thin-film samples where the growth of a new phase can be reaction-controlled. The latter process leads to flattening of concentration gradients in composition profiles which are then not representative anymore for the stability regions of the involved phases (see Fig. 1.3b)).

To extract information on the stability regions of the involved phases, Sm-1175-50h (Fig. 4.5d)) can be considered which shows a continuous composition change up to 95 at% Sm. This can be interpreted in terms of single-phase regions for the fluorite and bixbyite phases, although thermodynamic equilibrium may not have been attained after 50 h annealing. The composition, at which the transition between fluorite and bixbyite structure occurs, cannot be derived from this profile. The formation and growth of the phase with $I2_13$ structure (RL2) in Sm-1175-50h appears to be strongly reaction-limited as indicated by the extended composition plateau at 95 at% Sm. Strong reaction limitation is plausible because a substantial rearrangement of atoms is necessary at the interface between the monoclinic Sm_2O_3 and cubic $I2_13$ structure. After 100 h annealing (Sm-1175, Fig. 4.5e)), the plateau for the Sm-concentration suggests that the growth of RL1 into the CeO_2 substrate is also reaction-limited. Concentration plateaus are formed at 50 at% Sm for the bixbyite phase and 95 at% Sm for the phase with $I2_13$ structure. Reaction-limitation also dominates the growth of the reaction phases in Sm-1219 (Fig. 4.5f)) where the thickness of the RL1 and RL2 layers is substantially larger than the initial thickness of the Sm_2O_3 film. The same Sm-concentrations at the plateaus are found for Sm-1219 and Sm-1175.

Although definite statements on the stability regions of phases in the SDC system cannot be derived, some information was obtained which allows suggestions on the usage of the material as solid electrolyte. The low transition temperature to the monoclinic structure (987 °C) means that formation of monoclinic Sm_2O_3 or $Sm_xCe_{1-x}O_{2-x/2}$ during the fabrication process of the electrolyte must be avoided. This is especially important regarding the slow transformation kinetics from the monoclinic to the bixbyite structure. The absence of a miscibility gap between the fluorite and the bixbyite structure was not confirmed with certainty. This means that decomposition may take place. Decomposition into 2 different phases may result in the degradation of the material properties which are important for the device.

4.4.2 Cation Interdiffusion

Interdiffusion profiles were obtained by quantification of EELS and EDXS line scans perpendicular to the Sm_2O_3/CeO_2 interface. Composition profiles were not acquired close to grain boundaries in the Sm_2O_3 layer and CeO_2 substrate to avoid contributions from grain boundary diffusion. Only the samples Sm-987, Sm-1073, Sm-1122-40h, and Sm-1175-20h were investigated because these samples contain direct Sm_2O_3/CeO_2 interfaces. Specimens, which already showed continuous RLs, were not considered because the interdiffusion in these samples is superimposed by the growth of the RL into the CeO_2 substrate or the Sm_2O_3 film, respectively.

In general, composition-dependent interdiffusion coefficients cannot be excluded a priori which would lead to asymmetrical concentration profiles. The high symmetry of the profiles in Fig. 4.6 and the good fit of the experimental data by Eq. 1.8 indicate that composition dependence is not pronounced in the analyzed system under the applied conditions (temperature range and annealing time).

Assuming an Arrhenius-type relation, the temperature dependence of the diffusion coefficient can be described by Eq. 1.13 as introduced in Chapter 1.4.1. The activation enthalpy ΔH and frequency factor D_0 are derived from the Arrhenius diagram shown in Fig. 4.7, where D is plotted on a logarithmic scale as a function of the reciprocal absolute temperature $\frac{1}{T}$. Based on a linear regression, a straight line fits well to the determined interdiffusion coefficients. An activation enthalpy for bulk cation interdiffusion of $\Delta H = 2.69 \pm 0.31$ eV/atom and a frequency factor of $8.268 \cdot 10^{-11} \pm 2.48 \cdot 10^{-12}$ cm^{2}s^{-1} result from this fit, which characterize bulk cation interdiffusion across the interface. The errors result from the statistical error for the interdiffusion coefficients, the error of the linear regression and for the temperature determination. However, the good coincidence of the experimental data with the fit of a straight line with a relatively small error for H and D_0 supports the validity of the results despite the small number of data points.

A considerably larger activation enthalpy of 4.5 eV/atom was reported for grain growth experiments in cubic $Sm_{0.2}Ce_{0.8}O_{1.9}$ [131]. This indicates that cation transport across grain boundaries was impeded during grain growth compared to pure bulk cation interdiffusion. Grain coarsening in that study could have been also affected by the preparation route by oxalate coprecipitation which may have influenced the beginning of the sintering process. Such an effect can be excluded in our study.

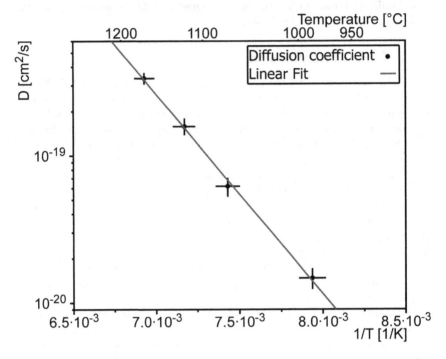

Figure 4.7: Arrhenius plot of the interdiffusion coefficients of the Sm samples with a linear fit curve to derive the activation enthalpy for interdiffusion

4.5 Comparison of the Phase Evolution of the SDC and GDC systems

Comparing the phase evolution of the $Sm_xCe_{1-x}O_{2-x/2}$ and $Gd_xCe_{1-x}O_{2-x/2}$ systems, the differences are remarkable. Neither the formation of the $I2_13$ structure nor broad Gd-concentration plateaus in interdiffusion profiles were observed in the GDC system (see Chapter 3.4.1). Instead, continuous interdiffusion profiles between the Gd_2O_3 thin film and the CeO_2 substrate without indication of a miscibility gap were found. The different behavior of the GDC system is attributed to the fact that Gd_2O_3 crystallizes in the cubic bixbyite structure in the investigated temperature range, which is similar to the fluorite structure of CeO_2.

In contrast, Sm_2O_3 crystallizes in the monoclinic phase. This requires heterogeneous nucleation of the bixbyite phase at the grain boundaries. First small RVs are observed before a complete RL with $Ia\overline{3}$ structure forms. Obviously a phase transformation must take place between the monoclinic and the cubic bixbyite structure at the Sm_2O_3/CeO_2 interface which is not neccessary at the Gd_2O_3/CeO_2 interface. Previous studies agree that the phase transformation from the monoclinic to the bixbyite structure is strongly affected by slow kinetics. Hence, interdiffusion and phase formation in the $Gd_xCe_{1-x}O_{2-x/2}$ system are less affected by slow phase-transformation kinetics compared to the $Sm_xCe_{1-x}O_{2-x/2}$ system.

Additionally, higher interdiffusion coefficients and a correspondingly lower activation enthalpy of 2.29 eV/atom were determined for bulk cation interdiffusion across the Gd_2O_3/CeO_2 interface in comparison to 2.69 ± 0.31 eV/atom for Sm_2O_3/CeO_2 interfaces. One can speculate that cation interdiffusion at the Gd_2O_3/CeO_2 interface is enhanced by the similarity of the crystal structures of CeO_2 and Gd_2O_3. The latter material occurs is the cubic bixbyite structure whereas Sm_2O_3 with monoclinic structure is present at the Sm_2O_3/CeO_2 interface for the investigated temperature range.

The lack of a miscibility gap between the fluorite and bixbyite structure and the indication that the bixbyite structure is the stable phase at low temperatures for $Gd_xCe_{1-x}O_{2-x/2}$ favors the use of this material system as a solid electrolyte in comparison to $Sm_xCe_{1-x}O_{2-x/2}$, where a miscibility gap at interesting material compositions cannot be excluded.

5 Summary

This work is concerned with phase formation and cation interdiffusion in the $Gd_xCe_{1-x}O_{2-x/2}$ and $Sm_xCe_{1-x}O_{2-x/2}$ systems at temperatures below 1200 °C. Diffusion-couple samples were fabricated and used in this work, which are well suited to analyze phase formation and to determine cation interdiffusion coefficients. The samples consist of CeO_2 substrates with a thin film of Sm_2O_3 and Gd_2O_3, respectively, deposited on the substrates. CeO_2 substrates with low porosity and impurity concentration were fabricated from high-purity CeO_2 powder. A multi-step procedure was applied involving Cold Isostatic Pressing (CIP), Hot Isostatic Pressing (HIP), polishing, and a final anneal to reduce the number of surface defects. Gd_2O_3 and Sm_2O_3 films with a thickness of about 150 nm and 180 nm were deposited by Pulsed Laser Deposition (PLD). This procedure resulted in thin films with amorphous structure on dense CeO_2 substrates with a planar substrate/film interface.

Transmission Electron Microscopy (TEM) was applied to study the structural properties. Bright-Field Transmission Electron Microscopy (BFTEM) images reveal the microstructure before and after annealing treatments. Crystal structure analyses were performed on the basis of Selected Area Electron Diffraction (SAED) patterns and Fourier-transformed High-Resolution Transmission Electron Microscopy (HRTEM) images, which were compared with simulated diffraction patterns. Cation interdiffusion coefficients were derived from cation-concentration profiles across the Gd_2O_3/CeO_2 and Sm_2O_3/CeO_2 interfaces obtained using Scanning Transmission Electron Microscopy (STEM) in combination with Energy-Dispersive X-ray Spectroscopy (EDXS) and Electron Energy Loss Spectroscopy (EELS). Both analytical techniques, EDXS and EELS, were applied as a consistency check of the measured compositions. EDXS and EELS line profiles were acquired from spectra taken in 1 nm intervals with an electron beam focused to a diameter of less than 1 nm, which yields a spatial resolution for composition analysis in the order of 1 nm depending on the local TEM sample thickness.

Studies of the Gd_2O_3-doped CeO_2 (GDC) system were carried out in the temperature range between 986 °C and 1266 °C. Annealing treatments were performed for up to 100 h in an argon/oxygen atmosphere. Reaction Layers (RLs) form between the thin Gd_2O_3 film and the CeO_2 substrate. Crystal structure analyses yield the fluorite structure for CeO_2 as expected. The bixbyite phase with $Ia\overline{3}$ structure was identified for the RLs and Gd_2O_3 at temperatures up to 1175 °C. This suggests that the transition from the bixbyite to the monoclinic structure for Gd_2O_3 occurs at higher temperatures in contrast to some other reports in literature. The composition profiles do not show any indication for a miscibility gap over the whole composition range of $Gd_xCe_{1-x}O_{2-x/2}$. Our results are consistent with literature data, which suggest a continuous transition between the fluorite and the bixbyite phases. Bulk interdiffusion coefficients for cation interdiffusion were determined for the $Gd_xCe_{1-x}O_{2-x/2}$ system from diffusion-couple samples. Interdiffusion coefficients were derived by fitting the diffusion-couple solution of Fick's second law to composition profiles across the Gd_2O_3/CeO_2-interface for temperatures between 986 °C and 1175 °C which yield cation interdiffusion coefficients of $6.53 \cdot 10^{-20} \pm 5.62 \cdot 10^{-21}$ cm^{2}s^{-1} at 986 °C, $1.63 \cdot 10^{-19} \pm 1.32 \cdot 10^{-20}$ cm^{2}s^{-1} at 1069 °C, $4.7 \cdot 10^{-19} \pm 4.6 \cdot 10^{-20}$ cm^{2}s^{-1} at 1122 °C, and $1.48 \cdot 10^{-19} \pm 0.12 \cdot 10^{-19}$ cm^{2}s^{-1} at 1175 °C. Exploiting the Arrhenius-type temperature dependence of the experimental interdiffusion coefficients, an activation enthalpy of 2.29 ± 0.22 eV/atom for the interdiffusion process and a temperature-independent frequency factor of $9.09 \cdot 10^{-11} \pm 4.18 \cdot 10^{-12}$ cm^{2}s^{-1} were obtained.

The Sm_2O_3-doped CeO_2 (SDC) system was studied at temperatures between 986 °C and 1270 °C. It shows differences compared to the GDC system with respect to reaction-phase formation. The formation of two reaction layers (RL1 and RL2) is observed depending on annealing time and temperatures. RL1 with the bixbyite structure (space group $Ia\overline{3}$) is first nucleated at grain boundaries in the Sm_2O_3 film and forms a complete layer after annealing for 50 h at 1175 °C. The crystal structure of RL2 on top of the RL1 was identified to be the cubic $I2_13$ structure. This phase is formed upon local excess of solubility limit of Ce (about 5 at%) in the monoclinic Sm_2O_3 film which demonstrates the low Ce-solubility in monoclinic Sm_2O_3. The $I2_13$ structure was previously only found for pure Sm_2O_3. However, the phase with $I2_13$ structure may be metastable at the temperatures applied in this study. Its formation could be favored by the presence of the cubic bixbyite with $Ia\overline{3}$ structure and minimization of strain in our thin-film samples and/or a non-equilibrium O-concentration. Another unexpected observation concerns broad plateaus in the composition profiles across the RLs, both showing almost constant Sm-concentrations after annealing for 100 h at 1175 °C and 1219 °C. Plateau formation can be understood by taking into account the growth kinetics of reaction phases in thin-film diffusion couples where the kinetics of reaction-phase formation dominates rather than interdiffusion. The flattened concentration gradients in the composition profiles are attributed to strongly reaction-controlled conditions, and the profiles are therefore not representative anymore for the stability regions of the involved phases. Bulk interdiffusion coefficients for cation interdiffusion were determined for the $Sm_xCe_{1-x}O_{2-x/2}$ system from diffusion couples for temperatures between 986 °C to 1175 °C. This yields diffusion coefficients of $1.47 \cdot 10^{-20} \pm 2.31 \cdot 10^{-21}$ cm^{2}s^{-1} for 987 °C, $6.17 \cdot 10^{-20} \pm 9.24 \cdot 10^{-21}$ cm^{2}s^{-1} for 1073 °C, $1.58 \cdot 10^{-19} \pm 2.03 \cdot 10^{-20}$ cm^{2}s^{-1} at 1122 °C, and $3.35 \cdot 10^{-19} \pm 3.10 \cdot 10^{-20}$ cm^{2}s^{-1} at 1175 °C. An activation enthalpy of 2.69 ± 0.31 eV/atom for the interdiffusion process and a temperature-independent frequency factor of $8.268 \cdot 10^{-11} \pm 2.48 \cdot 10^{-12}$ cm^{2}s^{-1} were determined.

There are remarkable differences in the phase evolution of the $Gd_xCe_{1-x}O_{2-x/2}$ system and the $Sm_xCe_{1-x}O_{2-x/2}$ system. Neither the formation of the $I2_13$ structure nor broad Gd-concentration plateaus in interdiffusion profiles were observed in the GDC system. The different behavior of the material systems is attributed to the fact that the Gd_2O_3 thin films crystallizes in the cubic bixbyite structure in the investigated temperature range which is similar to the fluorite structure of CeO_2. In contrast, the Sm_2O_3 thin film crystallizes in the monoclinic structure. Previous studies agree that phase transformation from the monoclinic to the bixbyite structure is strongly affected by slow kinetics. Hence, interdiffusion and phase formation in the $Gd_xCe_{1-x}O_{2-x/2}$ system are less affected by slow phase-transformation kinetics compared to the $Sm_xCe_{1-x}O_{2-x/2}$ system. Higher interdiffusion coefficients and a correspondingly lower activation enthalpy were determined for bulk interdiffusion across Gd_2O_3/CeO_2 interface in comparison to Sm_2O_3/CeO_2 interfaces. It is suggested that cation interdiffusion at the Gd_2O_3/CeO_2 interface is enhanced by the similarity of the crystal structures of CeO_2 and Gd_2O_3 which is not the case for the monoclinic Sm_2O_3.

In conclusion, this work expanded the knowledge and understanding of the $Gd_xCe_{1-x}O_{2-x/2}$ and $Sm_xCe_{1-x}O_{2-x/2}$ phase formation and yields information on phase diagrams towards lower temperatures. The lack of a miscibility gap in the GDC system makes this material particularly interesting for application in SOFCs. The material properties should not degrade due to decomposition. This is in contrast to Y_2O_3-doped ZrO_2 (YDZ) electrolytes which degrade due to the demixing into two different phases. In addition, bulk cation interdiffusion coefficients and their temperature dependence were determined quantitatively for the first time for Gd_2O_3/CeO_2 and Sm_2O_3/CeO_2 diffusion couples.

Bibliography

[1] W. Kollenberg, editor. *Technische Keramik (Technical Ceramics).* Vulkan-Verlag, Essen, 2nd edition, 2009.

[2] A. Heinzel, F. Mahlendorf, and J. Roes, editors. *Brennstoffzellen (Fuel Cells).* C. F. Müller Verlag, Heidelberg, 3rd edition, 2006.

[3] S. C. Singhal and K. Kendall. *High temperature solid oxide fuel cells: Fundamentals, Design and Applications.* Elsevier Ltd., New York, 1st edition, 2003.

[4] P. Holtappels and U. Stimming. Solid oxide fuel cells (SOFC). In W. Vielstich, A. Lamm, and H. A. Gasteiger, editors, *Handbook of Fuel Cells Vol.1: Fundamentals and Survey of Systems.* John Wiley & Sons Ltd, Chichester, 1st edition, 2003.

[5] A. Trovarelli. Catalytic properties of ceria and CeO_2-containing materials. *Catalysis Reviews: Science and Engineering*, 38:439–520, 1996.

[6] J. Kašpar, P. Fornasiere, and M. Graziani. Use of CeO_2-based oxides in three-way catalysis. *Catalysis Today*, 50:285–298, 1999.

[7] H.-J. Beie and A. Gnörich. Oxygen gas sensors based on CeO_2 thick and thin films. *Sensors and Actuators B*, 4:393–399, 1991.

[8] P. Jasinski, T. Suzuki, and H. U. Anderson. Nanocrystalline undoped ceria oxygen sensor. *Sensors and Actuators B: Chemical*, 95:73–77, 2003.

[9] S. Gupta, S. V. N. T. Kuchibhatla, M. H. Engelhard, V. Shutthanandan, P. Nachimuthu, W. Jiang, L. V. Saraf, S. Thevuthasan, and S. Prasad. Influence of samaria doping on the resistance of ceria thin films and its implications to the planar oxygen sensing devices. *Sensors and Actuators B: Chemical*, 139:380–386, 2009.

[10] N. Özer. Optical properties and electrochromic characterization of sol-gel deposited ceria films. *Solar Energy Materials & Solar Cells*, 68:391–400, 2001.

[11] I. Porqueras, C. Person, C. Corbella, M. Vives, E. Pinyol, and E. Bertran. Characteristics of e-beam deposited electrochromic CeO_2 thin films. *Solid State Ionics*, 165:131–137, 2003.

[12] R. Tarnuzzer, J. Colon, Patil S., and S. Seal. Vacancy engineered ceria nanostructures for protection from radiation-induced cellular damage. *Nano Letters*, 5:2573–2577, 2005.

[13] C. Lombardi and A. Mazzola. Exploiting the plutonium stockpiles in PWRs by using intert matrix fuel. *Annals of Nuclear Energy*, 23:1117–1126, 1996.

[14] H. Matzke, V. V. Rondinella, and T. Wiss. Materials research on inert matrices: a screening study. *Journal of Nuclear Materials*, 274:47–53, 1999.

[15] International Atomic Energy Agency (IAEA). Viability of inert matrix fuel in reducing plutonium amounts in reactors, 2006.

[16] N. G. Connelly, T. Damhus, R. M. Hartshorn, and A. T. Hutton. *Nomenclature of Inorganic Chemistry*. International Union of Pure and Applied Chemistry, Cambridge, 1st edition, 2005.

[17] T. Kudo and H. Obayashi. Oxygen ion conduction of the fluorite-type $Ce_{1-x}Ln_xO_{2-x/2}$ (ln = lanthanoid element). *Journal of the Electrochemical Society*, 122:142–147, 1975.

[18] R. T. Dirstine, R. N. Blumenthal, and T. F. Kuech. Ionic conductivity of calcia, yttria, and rare earth-doped cerium dioxide. *Journal of the Electrochemical Society*, 126:264–269, 1979.

[19] H. Yahiro, Y. Eguchi, K. Eguchi, and H. Arai. Oxygen ion conductivity of the ceria-samarium oxide system with fluorite structure. *Journal of Applied Electrochemistry*, 18:527–531, 1988.

[20] T. Inoue, T. Setoguchi, K. Eguchi, and H. Arai. Study of a solid oxide fuel cell with a ceria-based solid electrolyte. *Solid State Ionics*, 35:285–291, 1989.

[21] K. Yamashita, K. V. Ramanujachary, and M. Greenblatt. Hydrothermal synthesis and low temperature conduction properties of substituted ceria ceramics. *Solid State Ionics*, 81:53–60, 1995.

[22] G. B. Balazs and R. S. Glass. ac impedance studies of rare earth oxide doped ceria. *Solid State Ionics*, 76:155–162, 1995.

[23] W. Huang, P. Shuk, and M. Greenblatt. Properties of sol-gel prepared $Ce_{1-x}Sm_xO_{2-x/2}$ solid electrolytes. *Solid State Ionics*, 100:23–27, 1997.

[24] M. Mogensen, N. M. Sammes, and G. A. Tompsett. Physical, chemical and electrochemical properties of pure and doped ceria. *Solid State Ionics*, 129:63–94, 2000.

[25] H. Inaba and H. Tagawa. Ceria-based solid electrolytes. *Solid State Ionics*, 83:1–16, 1996.

[26] H. Yahiro, K. Eguchi, and H. Arai. Electrical properties and reducibilities of ceria-rare earth oxide systems and their application to oxide fuel cell. *Solid State Ionics*, 36:71–75, 1989.

[27] O. A. Marina and M. Mogensen. High-temperature conversion of methane on a composite gadolinia-doped ceria-gold anode. *Applied Catalysis A: General*, 189:117–126, 1999.

[28] O. A. Marina, C. Bagger, S. Primdahl, and M. Mogensen. A solid oxide fuel cell with a gadolinia-doped ceria anode: preparation and performance. *Solid State Ionics*, 123:199–208, 1999.

[29] C. Sun and U. Stimming. Recent anode advances in solid oxide fuel cells. *Journal of Power Sources*, 171:247–260, 2007.

[30] D. L. Maricle, T. E. Swarr, and S. Karavolis. Enhanced ceria - a low-temperature SOFC electrolyte. *Solid State Ionics*, 52:173–182, 1992.

[31] B. C. H. Steele. Appraisal of $Ce_{1-y}Gd_yO_{2-y/2}$ electrolytes for IT-SOFC operation at 500 °C. *Solid State Ionics*, 129:95–110, 2000.

[32] B. C. H. Steele and A. Heinzel. Materials for fuel-cell technologies. *Nature*, 414:345–352, 2001.

[33] E. Ryshkewitch. *Oxide Ceramics. Physical Chemistry and Technology*. Academic Press, London, 1st edition, 1960.

[34] K. K. Srivastava, R. N. Patil, C. B. Choudhary, G. K. Gokhale, and E. C. Subbarao. Revised phase diagram of the system ZrO_2-$YO_{1.5}$. *British Ceramic: Transactions and Journal*, 73:85–91, 1974.

[35] R. C. Garvie, R. H. Hannik, and R. T. Pascoe. Ceramic steel? *Nature*, 1975:703–704, 1975.

[36] A. H. Heuer, R. Chaim, and V. Lanteri. Review: Phase transformations and microstructural characterization of alloys in the system Y_2O_3-ZrO_2. In *Advances in Ceramics: Science and Technology of Zirconia III*, volume 24, pages 3–20. American Ceramics Society, Ohio, 1988.

[37] M. Rühle. Microscopy of structural ceramics. *Advanced Materials*, 9:195–217, 1997.

[38] O. Yamamoto, Y. Arachi, H. Sakai, Y. Takeda, N. Imanishi, Y. Mizutani, M. Kawai, and Y. Nakammura. Zirconia based oxide ion conductores for solid oxide fuel cells. *Ionics*, 4:403–408, 1998.

[39] H. G. Scott. Phase relationships in the zirconia-yttria system. *Journal of Materials Science*, 10:1527–1535, 1975.

[40] A. Weber and E. Ivers-Tiffée. Materials and concepts for solid oxide fuel cells (SOFCs) in stationary and mobile applications. *Journal of Power Sources*, 127:273–283, 2004.

[41] J. Kondoh, T. Kawashima, S. Kikuchi, Y. Tomii, and Y. Ito. Effect of aging on yttria-stabilized zirconia. *Journal of the Electrochemical Society*, 145:1527–1536, 1998.

[42] C. C. Appel, N. Bonanos, A. Horsewell, and S. Linderoth. Ageing behaviour if zirconia stabilised by yttria and manganese oxide. *Journal of Materials Science*, 36:4493–4501, 2001.

[43] N. Balakrishnan, T. Takeuchi, K. Nomura, H. Kageyama, and Y. Takeda. Aging effect of 8 mol% YSZ ceramics with different microstructures. *Journal of the Electrochemical Society*, 151:A1286–A1291, 2004.

[44] M. Hattori, Y. Takeda, Y. Sakaki, A. Nakanishi, S. Ohara, K. Mukai, J.-H. Lee, and T. Fukui. Effect of aging on conductivity of yttria stabilized zirconia. *Journal of Power Sources*, 126:23–27, 2004.

[45] M. Hattori, Y. Takeda, J.-H. Lee, S. Ohara, K. Mukai, T. Fukui, S. Takahashi, Y. Sakaki, and A. Nakanishi. Effect of annealing on the electrical conductivity of the Y_2O_3-ZrO_2 system. *Journal of Power Sources*, 131:247–250, 2004.

[46] B. Butz, P. Kruse, H. Störmer, D. Gerthsen, A. Müller, A. Weber, and E. Ivers-Tiffée. Correlation between microstructure and degradation in conductivity for cubic Y_2O_3-doped ZrO_2. *Solid State Ionics*, 177:3275–3284, 2006.

[47] B. Butz, R. Schneider, D. Gerthsen, M. Schowalter, and A. Rosenauer. Decomposition of 8.5 mol.% Y_2O_3-doped zirconia and its contribution to the degradation of ionic conductivity. *Acta Materialia*, 57:5480–5490, 2009.

[48] K. Eguchi. Ceramic materials containing rare earth oxides for solid oxide fuel cell. *Journal of Alloys and Compounds*, 250:486–491, 1997.

[49] S. P. Simner, M. D. Anderson, M. H. Engelhard, and J. W. Stevenson. Degradation mechanisms of La-Sr-Co-FeO_3 SOFC cathodes. *Electrochemical and Solid-State Letters*, 9:A478–A481, 2006.

[50] H.-G. Jung, Y.-K. Sun, H. Y. Jung, J. S. Park, H.-R. Kim, G.-H. Kim, H.-W. Lee, and J.-H. Lee. Investigation of anode-supported SOFC with cobalt-containing cathode and GDC interlayer. *Solid State Ionics*, 179:1535–1539, 2008.

[51] N. Jordan, S. Uhlenbruck, V. A. C. Haanappel, H. P. Buchkremer, D. Stöver, and W. Mader. $Ce_{0.8}Gd_{0.2}O_{2-d}$ protecting layers manufactured by physical vapor deposition for IT-SOFC. *Solid State Ionics*, 179:919–923, 2008.

[52] H. Chang, D. Tsai, W. Chung, Y. Huang, and M. Le. A ceria layer as diffusion barrier between LAMOX and lanthanum strontium cobalt ferrite along with the impedance analysis. *Solid State Ionics*, 180:412–417, 2009.

[53] C. Sun, R. Hui, and J. Roller. Cathode materials for solid oxide fuel cells: areview. *Journal of Solid State Electrochemistry*, 14:1125–1144, 2009.

[54] L. Eyring. The binary rare earth oxides. In K. A Gschneider, Jr. and L. Eyring, editors, *Handbook on the Physics and Chemistry of Rare Earths Vol. 3*, pages 337–399. North-Holland Publishing Company, Amsterdam, New York, Oxford, 1st edition, 1979.

[55] G. Adachi and N. Imanaka. The binary rare earth oxides. *Chemical Reviews*, 98:1479–1514, 1998.

[56] M. Zinkevich. Thermodynamics of rare earth sesquioxides. *Progress in Material Science*, 52:597–647, 2006.

[57] D. J. M. Bevan and E. Summerville. Mixed rare earth oxides. In K. A Gschneider, Jr. and L. Eyring, editors, *Handbook on the Physics and Chemistry of Rare Earths Vol. 3*, pages 401–524. North-Holland Publishing Company, Amsterdam, New York, Oxford, 1st edition, 1979.

[58] R. W. G. Wyckoff. *Crystal Structures Vol. 2*. John Wiley & Sons, Inc., 2nd edition, 1964.

[59] A. F. Wells. *Structural inorganic chemistry*. Clarendon Press (Oxford University Press), Oxford, 5th edition, 1984.

[60] C. E. Housecroft and A. G. Sharpe. *Anorganische Chemie (Inorganic Chemistry)*. Pearson Studium, München, 2nd edition, 2006.

[61] B. Wu, M. Zinkevich, F. Aldinger, D. Wen, and L. Chen. Ab initio study on structure and phase transition of A- and B-type rare-earth sesquioxides Ln_2O_3 (Ln = La-Lu, Y, and Sc) based on density function theory. *Journal of Solid State Chemistry*, 180:3280–3287, 2007.

[62] R. D. Shannon. Revised effective ionic radii and systematic studies of interatomich distances in halides and chacogenides. *Acta Crystallographica A*, 32:751–767, 1976.

[63] P. P. Ewald and C. Hermann. *Strukturbericht 1913–1926 (Structure Report 1913–1926)*. Akademische Verlagsgesellschaft M.B.H., Leipzig, 1st edition, 1931.

[64] T. Hahn, editor. *International Tables for Crystallography Volume A Space-group Symmetry*. The International Union of Crystallography, Dordrecht, Boston, London, 5th edition, 2002.

[65] V. M. Goldschmidt, F. Ulrich, and T. Barth. Geochemische Verteilungsgesetze der Elemente IV: Zur Kristallstruktur der Oxyde der seltenen Erdmetalle (Laws on the geochemical distribution of the elements IV: On the crystal structure of the oxides of the rare earths). *Skrifter utgitt av det Norske Videnskaps-Akademi i Oslo 1: Matematisk-Naturvidenskapelig Klasse 1925*, 5:1–24, 1925.

[66] V. M. Goldschmidt and L. Thomassen. Die Kristallstruktur natürlicher und synthetischer Oxyde von Uran, Thorium und Cerium (the crystal structure of natural and synthetic oxides of Uranium, Thorium, and Cerium). *Skrifter utgitt av det Norske Videnskaps-Akademi i Oslo 1: Matematisk-Naturvidenskapelig Klasse 1923*, 2:1–49, 1923.

[67] G. Brauer and H. Gradinger. Über heterotype Mischphasen bei Seltenerdoxyden II: Die Oxydsysteme des Cers und des Praesodyms (On (heterotype) mixed phases of rare earth oxides II: The oxide systems of Cerium and Paseodymium). *Zeitschrift für anorganische und allgemeine Chemie*, 277:89–95, 1954.

[68] G. Brauer, K. A. Gingerich, and U. Holtschmidt. Über die Oxyde des Cers - IV: Die Sauerstoffzersetzungsdrucke im System der Ceroxyde (On the oxides of Cerium - IV: The oxide pressures of decomposition in the systems of the (different) cerium oxides). *Journal of Inorganic & Nuclear Chemistry*, 16:77–86, 1960.

[69] G. Brauer and K. A. Gingerich. Über die Oxyde des Cers - V: Hochtemperaturröntgenuntersuchungen an Ceroxyden (On the oxides of Cerium - V: High-temperature X-ray analysis of Cerium oxides). *Journal of Inorganic & Nuclear Chemistry*, 16:87–99, 1960.

[70] M. Ricken, J. Nölting, and I. Riess. Specific heat and phase diagram of nonstoichiometric ceria. *Journal of Solid State Chemistry*, 54:89–99, 1984.

[71] K. Yasunaga, K. Yasuda, S. Matsamura, and T. Sonoda. Nucelation and growth of defect clusters in CeO_2 irradiated with electrons. *Nuclear Instruments and Methods in Physics Research B - Beam Interactions with Materials and Atoms*, 250:114–118, 2006.

[72] G. Brauer and H. Gradinger. Über heterotype Mischphasen bei Seltenerdoxyden I (On (heterotype) mixed phases of rare earth oxides I). *Zeitschrift für anorganische und allgemeine Chemie*, 276:209–348, 1954.

[73] A. Bartos, K. P. Lieb, M. Uhrmacher, and D. Wiarda. Refinement of atomic positions in bixbyite oxides using perturbed angular correlation spectroscopy. *Acta Crystallographica B*, 49:165–169, 1993.

[74] V. Grover, S. N. Achary, and A. K. Tyagi. Structural analysis of excess-anion C-type rare earth oxide: a case study with $Gd_{1-x}Ce_xO_{1.5+x/2}$ (x=0.20 and 0.40). *Journal of Applied Crystallography*, 36:1082–1084, 2003.

[75] A. A. Zav'Yalova, R. M. Imamov, N. A. Ragimli, and S. A. Semiletov. Electron-diffraction study of the structure of cubic $C\text{-}Sm_2O_3$. *Kristallografiya*, 21:727–729, 1976.

[76] D. T. Cromer. The crystal structure of monoclinic Sm_2O_3. *Journal of Physical Chemistry*, 61:753–755, June 1957.

[77] F. Ye, T. Mori, D. R. Ou, and J. Drennan. A structure model of nano-sized domain in Gd-doped ceria. *Solid State Ionics*, 180:1414–1420, 2009.

[78] M. Foex and J.-P. Traverse. Remarques sur les tranformations cristallines présentées a haute température par les sesquioxydes de terres rares (Study on the crystal structure transformations at high temperatures of the rare

earth sesquioxides). *Revue Internationale des Hautes Temperatures et des Réfractaires*, 122:429–453, 1966.

[79] J.-P. Coutures, R. Verges, and M. Foex. Valeurs comparées des temperatures de solidification des différent sesquioxydes de terres rares; influence de l'atmosphère (Comparison of the solidification temperatures of different sesquioxides; influence of the atmosphere). *Revue Internationale des Hautes Temperatures et des Réfractaires*, 12:181–185, 1975.

[80] A.V. Shevthenko and L.M. Lopato. TA method application to the highest refractory oxide systems investigation. *Thermochimia Acta*, 93:537–540, 1985.

[81] W. H. Zachariasen. The crystal structure of the modification C of the sesquioxides of the rare earth metals, and of Indium and Thallium. *Norsk Geologisk Tidsskrift*, 9:310–316, February 1927.

[82] W. H. Zachariasen. Untersuchungen ueber die Kristallstruktur von Sesquioxyden und Verbindungen ABO_3 (Investigations of the crystal structure of the sesquioxides and the ABO_3 compounds). *Skrifter utgitt av det Norske Videnskaps-Akademi i Oslo 1: Matematisk-Natuvidenskapelig Klasse 1928*, 4:1–165, 1928.

[83] L. Pauling and M. D. Shappell. The crystal structure of bixbyite and the C-modification of the sesquioxides. *Zeitschrift für Kristallographie*, 75:128–142, 1930.

[84] H. Bommer. Die Gitterkonstanten der C-Formen der Oxyde der seltenen Erdmetalle (The lattice constants of the C-forms of the oxides of the rare earths). *Zeitschrift für anorganische und allgemeine Chemie*, 241:273–280, 1939.

[85] D. H. Templeton and C. H. Dauben. Lattice parameters of some rare earth compounds and a set of crystal radii. *Journal of the American Chemical Society*, 76:5237–5239, 1954.

[86] N. Hirosaki, S. Ogata, and C. Kocer. Ab initio calculation of the crystal structure of the lanthanide Ln_2O_3 sesquioxides. *Journal of Alloys and Compounds*, 351:31–34, 2003.

[87] O. J. Guentert and R. L. Mozzi. The monoclinic modification of gadolinium sesquioxide Gd_2O_3. *Acta Crystallographica*, 11:746, 1958.

[88] K. I. Portnoi, V. I. Fadeeva, and N. I. Timofeeva. The polymorphism of some rare earth oxides and their reaction with water. *Atomnaya Energiya*, 14:559–552, 1963.

[89] C. E. Curtis and J. R. Johnson. Ceramic properties of Samarium oxide and Gadolinium oxide; X-ray studies of other rare-earth oxides and some compounds. *Journal of the American Ceramic Society*, 40:15–19, 1957.

[90] G. L. Ploetz, C. W. Krystyniak, and H. E. Dumas. Sintering characteristics of rare-earth oxides. *Journal of the American Ceramic Society*, 41:551–554, 1958.

[91] R. S. Roth and S. J. Schneider. Phase equilibria in systems involving the rare-earth oxides. Part I. Polymorphism of the oxides of the trivalent rare-earth ions. *Journal of Research of the National Bureau of Standards - A. Physics and Chemistry*, 64A:309–316, April 1960.

[92] M. W. Shafer and R. Roy. Rare-earth polymophism and phase equilibria in rare-earth oxide-water systems. *Journal of the American Ceramic Society*, 42:563–570, 1959.

[93] I. Warshaw and R. Roy. Polymorphism of the rare earth sesquioxides. *Journal of Physical Chemistry*, 65:2048–2051, 1961.

[94] G. Brauer and R. Müller. Beiträge zur Polymorphie der Sesquioxide der Seltenen Erden (Contributions to the polymorphisms of the rare earths). *Zeitschrift für anorganische und allgemeine Chemie*, 321:234–245, 1963.

[95] H. R. Hoekstra. Phase relationships in the rare earth sesquioxides at high pressure. *Inorganic Chemistry*, 5:754–757, December 1965.

[96] F. X. Zhang, M. Lang, J. W. Wang, U. Becker, and R. C. Ewing. Structural phase transitions of cubic Gd_2O_3 at high pressures. *Physical Review B*, 78:1–9, 2008.

[97] M. Zinkevich, D. Djurovic, and Fritz Aldinger. Thermodynamics of rare earth doped ceria. In *Proceedings of the 7th European Solid Oxide Fuel Cells Forum*, 2006.

[98] V. Grover and A. K. Tyagi. Phase relations, lattice thermal expansion in the CeO_2-Gd_2O_3 system, and stabilization of cubic gadolinia. *Materials Research Bulletin*, 39:859–866, 2004.

[99] A. Artini, G. A. Costa, M. Pani, A. Lausi, and J. Plaisier. Structural characterization of the CeO_2/Gd_2O_3 mixed system by synchrotron X-ray diffraction. *Journal of Solid State Chemistry*, 190:24–28, 2012.

[100] F. Ye, T. Mori, D. R. Ou, J. Zou, G. Auchterlonie, and J. Drennan. Compositional and structural characteristics of nano-sized domains in Gadolinium-doped ceria. *Solid State Ionics*, 179:827–831, 2008.

[101] Z. Li, T. Mori, F. Ye, D. R. Ou, J. Zou, and J. Drennan. Ordered structures of defect clusters in Gadolinium-doped ceria. *The Journal of Chemical Physics*, 134:1–7, 2011.

[102] F. Ye, D. R. Ou, and T. Mori. Microstructural evolution in a CeO_2-Gd_2O_3 system. *Microscopy and Microanalysis*, 18:162–170, 2012.

[103] T. Mori and J. Drennan. Influence of microstructure on oxide ionic conductivity in doped CeO_2 electrolytes. *Journal of Electroceramics*, 17:749–757, 2006.

[104] D. J. M. Bevan, W. W. Barker, and T. C. Parks. Mixed oxides of the type MO_2(fluorite)-M_2O_3. Part 2: Non-stoichiometry in ternary rare-earth oxide systems. *Proceedings of the Fourth Conference on Rare Earth Research*, pages 441–468, 1964.

[105] W. B. White and V. G. Keramidas. Vibrational spectra of oxides with the C-type rare earth oxide structure. *Spectrochimia Acta*, 28A:501–509, 1971.

[106] S. A. Semiletov, R. M. Imamov, and N. A. Ragimli. Obtaining thin films of Gd, Sm, and Eu oxides and their structural examination by electron diffraction. *Russian metallurgy*, 6:60–64, 1975.

[107] S. A. Semiletov, R. M. Imamov, N.A. Ragimli, and L. I. Man. Crystal structure of thin films of some rare earth oxides. *Thin Solid Films*, 32:325–328, 1976.

[108] R. M. Douglass and Eugene Staritzky. Samarium sesquioxide, Sm_2O_3, form B. *Analytical Chemistry*, 28:552, 1956.

[109] P. C. Boulesteix, B. Pardo, P. E. Caro, M. Gasgnier, and C. Henry la Blanchetais. Etude de couches minces de sesquioxyde de samarium type B par microscopie et diffraction electroniques (Study on thin layers of the B-form of samarium sesquioxide using microscopy and electron diffraction). *Acta Crystallographica B*, 27:216–219, March 1971.

[110] T. Schleid and G. Meyer. Single crystals of rare earth oxides from reducing halide melts. *Journal of the Less-Common Metals*, 149:73–80, 1989.

[111] B. J. Kennedy and M. Avdeev. The structure of B-type $Sm_2O_3 \cdot$ a powder neutron diffraction study using enriched [154]Sm. *Solid State Sciences*, 13:1701–1703, June 2011.

[112] T. Atou, K. Kusaba, Y. Tsuchida, W. Utsumi, T. Yagi, and Y. Syono. Reversible B-type - A-type transition of Sm_2O_3 under high pressure. *Materials Research Bulletin*, 24:1171–1176, 1989.

[113] M. R. Esquivel, A. E. Bohé, and D. M. Pasquevich. Synthesis of samarium sesquioxide from thermal decomposition of Samarium oxychloride. *Materials Research Bulletin*, 42:553–562, 2007.

[114] B. P. Mandal, V. Grover, and A. K. Tyagi. Phase relations, lattice thermal expansion in $Ce_{1-x}Eu_xO_{2-x/2}$ and $Ce_{1-x}Sm_xO_{2-x/2}$ systems and stabilization of cubic REO_3 (RE: Eu, Sm). *Materials Science and Engineering A*, 430:120–124, May 2006.

[115] T. C. Parks and D. J. M. Bevan. Mixed oxides of the type MO_2(fluorite)-$MO_{1.5}$ VI: phase relationships in the systems CeO_2-$SmO_{1.5}$ and CeO_2-$NdO_{1.5}$. *Revue de Chimie minérale*, 10:15–28, September 1973.

[116] H. Mehrer. *Diffusion in Solids*. Springer, Berlin Heidelberg, 1st edition, 2007.

[117] D. A. Porter and K. E. Easterling. *Phase Transformations in Metals and Alloys*. Nelson Thornes Ltd, Cheltenham, 2nd edition, 2001.

[118] A. Fick. Ueber diffusion. *Annalen der Physik und Chemie*, 94:59–86, 1855.

[119] U. Gösele and K. N. Tu. Growth kinetics of planar binary diffusion couples: "Thin-film case" versus "Bulk cases". *Journal of Applied Physics*, 53:3252–3260, 1981.

[120] V. I. Dybkov. *Reaction diffusion and solid state chemical kinetics*. Trans Tech Publications Ltd, Stafa-Zuerich, Switzerland, 2nd edition, 2010.

[121] J. E. Burke. Some factors affecting the rate of grain growth in metals. *Transactions of the Americab Institute of Mining and Metallurgical Engineers*, 180:73–91, 1949.

[122] J. E. Burke and D. Turnbull. Recrystallization and grain growth. *Progress in metal physics*, 3:220–292, 1952.

[123] M. F. Yan, R. M. Cannon, and H. K. Bowen. Grain boundary migration in ceramics. In R. M. Fulrath and J. A. Pask, editors, *Ceramic Microstructures '76*, pages 276–307. Westview Press, Boulder, Colorado, 1976.

[124] J. T. Hansen, R. P. Rusin, M.-H. Teng, and D. L. Johnsons. Combined-stage sintering model. *Journal of the American Ceramic Society*, 75:1129–1135, 1992.

[125] C. Herring. Effects of change of scale on sintering phenomena. *Journal of Applied Physics*, 21:301–303, 1950.

[126] E. Jud, C. B. Huwiler, and L. J. Gauckler. Sintering analysis of undoped and Cobalt oxide doped ceria solid solutions. *Journal of the American Ceramic Society*, 88:3013–3019, 2005.

[127] J. Liang, Q. Zhu, Z. Xie, W. Huang, and C. Hu. Low-temperature sintering behaviours of nanosized $Ce_{0.8}Gd_{0.2}O_{1.9}$ powder synthesized by co-precipitation combined with supercritical drying. *Journal of Power Sources*, 194:640–645, 2009.

[128] W. S. Young and I. B. Cutler. Initial sintering with constant rates of heating. *Journal of the American Ceramic Society*, 53:659–663, 1970.

[129] V. Gil, C. Moure, P. Duran, and J. Tartaj. Low-temperature densification and grain growth of Bi_2O_3-doped-ceria gadolinia ceramics. *Solid State Ionics*, 178:359–365, 2007.

[130] J. L. M. Rupp, A. Infortuna, and L. J. Gauckler. Microstrain and self-limited grain growth in nanocrystalline ceria ceramics. *Acta Materialia*, 54:1721–1730, 2006.

[131] Y. Okawa, T. Matsumoto, T. Doi, and Y. Hirata. Thermal stability of nanometer-sized NiO and Sm-doped ceria powders. *Journal of Materials Research*, 17:2266–2274, 2002.

[132] R. Yan, F. Chu, Q. Ma, X. Liu, and G. Meng. Sintering kinetics of Samarium doped ceria with addition of Cobalt oxide. *Materials Letters*, 60:3605–3609, 2006.

[133] D. B. Williams and C. B. Carter. *Transmission Electron Microscopy - A textbook for Materials Science.* Springer Science+Business Media, New York, 2nd edition, 2009.

[134] L. Reimer and H. Kohl. *Transmission Electron Microscopy - Physics of Image Formation.* Springer Science+Business Media, New York, 5th edition, 2008.

[135] L. Reimer. *Scanning Electron Microscopy: Physics of Image Formation and Microanalysis.* Springer, Berlin Heidelberg New York, 2nd edition, 1998.

[136] P. A. Stadelmann. EMS - a software package for electron diffraction analysis and HREM image simulation in materials science. *Ultramicroscopy*, 21:131–146, 1987.

[137] M. Knoll and E. Ruska. Das Elektronenmikroskop (The electron microscope). *Zeitschrift für Physik*, 78:318–339, 1932.

[138] H. H. Rose. *Geometrical Charged-Particle Optics.* Springer, Berlin Heidelberg, 1st edition, 2009.

[139] M. von Ardenne. Das Elektronen-Rastermikroskop. Praktische Ausführung. (The scanning electron microscope. Practical application.). *Zeitschrift für technische Physik*, 19:407–416, 1938.

[140] M. von Ardenne. Das Elektronen-rastermikroskop. Theoretische Grundlagen. (The scanning electron miscroscope. Theoretical considerations.). *Zeitschrift für Physik*, 109:553–572, 1938.

[141] S. J. Pennycook and P. D. Nellist, editors. *Scanning Transmission Electron Microscopy Imaging an Analysis.* Springer Science+Business Media, New York, Dordrecht , Heidelberg, London, 1st edition, 2011.

[142] P. Müller. *Electron Microscopical Investigations of Doped and Undoped $Ba_{0.5}Sr_{0.5}Co_{0.8}Fe_{0.2}O_{3-d}$ for Oxygen Separation Membranes.* PhD thesis, Karlsruher Institut für Technologie (KIT), Karlsruhe, 2013.

[143] I. P. Jones. *Chemical Microanalysis using electron beams.* The Institute of materials, London, 1st edition, 1992.

[144] J. Hillier and R. F. Baker. Microanalysis by means of electrons. *Journal of Applied Physics*, 15:663–675, 1944.

[145] R. Castaing. *Application des sondes électroniques a une méthode d'analyse ponctuelle chimique et cristallographique (Application of an electron probe to analyse the local chemical composition and crystal structure).* PhD thesis, Offices national d'études et de recherches aérospatiales, Paris, 1951.

[146] G. Cliff and G. W. Lorimer. The quantitative analysis of thin specimens. *Journal of Microscopy*, 103:203–207, 1975.

[147] J. A. Bearden. X-ray wavelenghths. *Reviews of Modern Physics*, 39:78–124, 1967.

[148] E. W. White and G. G. Johnson, Jr. *X-Ray and Absorption Wavelengths and Two-Theta Tables.* Number DS 37A in ASTM Data Series. American Society for Testing and Materials, Philadephia, 2nd edition, 1970.

[149] P. Bougouer. *Essai d'Optique, sur la gradation de la lumière (Essay on optics, on the gradation of light).* Avec Approbation et Privilege du Roi, Paris, 1729.

[150] J. H. Lambert. *Photometrie (Photometria Sive De Mensura Et Gradibus Luminis, Colorum Et Umbrae).* Verlag von Wilhelm Engelmann, Leipzig, 1760.

[151] J. H. Hubbell. Photon mass attenuation and energy-absorption coefficients from 1 keV to 20 MeV. *International Journal of Applied Radiation and Isotopes*, 33:1269–1290, 1982.

[152] S. M. Seltzer. Calculation of photon mass energy-transfer and mass energy-absorption coefficients. *Radiation Research*, 136:147–170, 1993.

[153] C. T. Chantler. Theoretical form factor, attenuation and scattering tabulation for Z = 1-92 from E = 1-10 eV to E = 0.4-1.0 MeV. *Journal of Physical and Chemical Reference Data*, 24:71–643, 1995.

[154] C. T. Chantler. Detailed tabulation of atomic form factors, photoelectric absorption and scattering cross section, and mass attenuation coefficients in the vicinity of absorption edges in the soft X-ray (Z = 30-36, Z = 60-89, E = 0.1 keV-10 keV), addressing convergence issues of earlier work. *Journal of Physical and Chemical Reference Data*, 29:597–1048, 2000.

[155] C. T. Chantler, K. Olsen, R. A. Dragoset, J. Chang, A. R. Kishore, S. A. Kotochigova, and D. S. Zucker. X-ray form factor, attenuation, and scattering tables, 2005.

[156] E. Rudberg. *Soft X-rays and secondary electrons: Velocity measurements on the electron emission produced by soft X-rays and by electron impact.* PhD thesis, Stockholm universitet (Stockholm university), Stockholm, 1929.

[157] G. Ruthemann. Diskrete Energieverluste schneller Elektronen in Festkörpern (Discrete energy losses of fast electrons in solids). *Naturwissenschaften*, 29:648, 1941.

[158] R. F. Egerton. *Electron Energy-Loss Spectroscopy in the Electron Microscope.* Plenum Press, New York, London, 2nd edition, 1996.

[159] R. Brydson. *Electron Energy Loss Spectroscopy.* Number 48 in Microscopy Handbooks. BIOS Scientific Publishers Limited, Oxford, 1st edition, 2001.

[160] C. C. Ahn, editor. *Transmission Electron Energy Loss Spectrometry in Materials Science an the EELS ATLAS.* Wiley-VCH Verlag GmbH Co. KGaA, Weinheim, 2nd edition, 2004.

[161] T. Malis, S. C. Cheng, and R. F. Egerton. EELS log-ratio technique for specimen-thickness measurement in the TEM. *Journal of Electron Microscopy Technique*, 8:193–200, 1988.

[162] R. F. Egerton. K-shell ionization cross-sections for use in microanalysis. *Ultramicroscopy*, 4:169–179, 1979.

[163] R. D. Leapman, P. Rez, and D. F. Mayers. K, L, and M shell generalized oscillator strengths and ionization cross sections for fast electron collisions. *The Journal of Chemical Physics*, 72:1232–1243, 1980.

[164] P. Rez. Cross-sections for energy loss spectrometry. *Ultramicroscopy*, 9:283–288, 1982.

[165] C. C. Ahn and P. Rez. Inner shell edge profiles in electron energy loss spectrometry. *Ultramicroscopy*, 17:105–116, 1985.

[166] L. Dieterle, B. Butz, and E. Müller. Optimized Ar^+-ion miling procedure for TEM cross-section sample preparation. *Ultramicroscopy*, 111:1636–1644, 2011.

Own Publications and Contributions to Conferences

Reviewed Publications

- C. Rockenhäuser, B. Butz, N. Schichtel, J. Janek, R. Oberacker, M. J. Hoffmann, and D. Gerthsen. Microstructure evolution and cation interdiffusion in thin Gd_2O_3 films on CeO_2 substrates. *Journal of the European Ceramic Society* **34**, 1235-1242 (2014).

- C. Rockenhäuser, B. Butz, H. Störmer, and D. Gerthsen. Microstructure evolution and cation interdiffusion in thin Sm_2O_3 films on CeO_2 substrates. *Journal of the American Ceramic Society* **97**, 3276-3285 (2014).

Contributions to Conferences

Oral Presentations

- C. Rockenhäuser, B. Butz, N. Schichtel, C. Korte, J. Janek, and D. Gerthsen. Cation transport studies by Samarium (Sm) dopant diffusion into polycrystalline CeO_2 substrates. *1st International Conference on Materials for Energy*, Karlsruhe, Germany, July 4-8 (2010).

- C. Rockenhäuser, B. Butz, D. Gerthsen, N. Schichtel, C. Korte, J. Janek. Cation transport studies of Samarium (Sm) dopant diffusion into polycrystalline CeO_2. *Electroceramics XII*, Trondheim, Norway, August 14-17 (2010).

- C. Rockenhäuser, B. Butz, and D. Gerthsen. Cation diffusion and phase formation at Sm_2O_3/CeO_2 and Gd_2O_3/CeO_2 interfaces. *Microscopy Conference (MC) 2011*, Regensburg, Germany, August 25-30 (2013).

Poster Presentations

- C. Rockenhäuser, B. Butz, N. Schichtel, C. Korte, J. Janek, and D. Gerthsen. Cation transport studies by dopant (Sm, Gd) diffusion in polycrystalline CeO_2 substrates. *Microscopy Conference (MC) 2009*, Graz, Austria, August 30-September 4 (2009).

- C. Rockenhäuser, B. Butz, and D. Gerthsen. Cation interdiffusion and phase formation studies from donor thin films (Gd, Sm) into polycrystalline CeO_2 substrates *Microscopy Conference (MC) 2011*, Kiel, Germany, August 28-September 2 (2011).

- C. Rockenhäuser, B. Butz, and D. Gerthsen. Cation diffusion and phase formation at Gd_2O_3/CeO_2 interfaces. *The 15th European Microscopy Congress*, Manchester, United Kingdom, September 16-21 (2012).